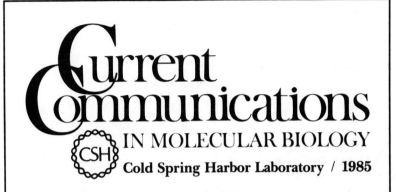

IN MOLECULAR BIOLOGY
Cold Spring Harbor Laboratory / 1985

Eukaryotic Transcription

The Role of *cis*- and *trans*-acting Elements in Initiation

Edited by

Yakov Gluzman

Cold Spring Harbor Laboratory

Titles in
Current Communications in Molecular Biology
PLANT INFECTIOUS AGENTS
ENHANCERS AND EUKARYOTIC GENE EXPRESSION
PROTEIN TRANSPORT AND SECRETION
IMMUNE RECOGNITION OF PROTEIN ANTIGENS
EUKARYOTIC TRANSCRIPTION

EUKARYOTIC TRANSCRIPTION
The Role of *cis*- and *trans*-acting
Elements in Initiation
© 1985 by Cold Spring Harbor Laboratory
All rights reserved
International Standard Book Number 0-87969-186-7
Book design by Emily Harste
Printed in the United States of America

1260 2670

Cover: Interactions between the SV40 early promoter elements and multiple *trans*-acting cellular proteins. Figure shows the DNase I footprints (vertical bars with open circles, shaded areas) of the proteins of a HeLa cell nuclear extract that interacts with the late (*left*) and the early (*right*) coding strands of the 21-bp repeat region (SV40 coordinates 40–100) and the enhancer element containing a single 72-bp sequence (from coordinate 107 to the *Pvu*II site). The sites that become hypersensitive to DNase I upon protein binding are indicated by a vertical bar and a filled circle on the left of the footprints and by arrows on the diagrams. (For further information, see A. Wildeman et al., this volume). (Figure courtesy of P. Chambon.)

The individual summaries contained herein should not be treated as publications or listed in bibliographies. Information contained herein can be cited as personal communication contingent upon obtaining the consent of the author. The collected work may, however, be cited as a general source of information on this topic.

All Cold Spring Harbor Laboratory publications may be ordered directly from Cold Spring Harbor Laboratory, Box 100, Cold Spring Harbor, New York 11724. (Phone: 1-800-843-4388). In New York State (516) 367-8425.

151704

Conference Participants

Kim Arndt, Whitehead Institute for Biomedical Research, Cambridge, Massachusetts

Arnold Berk, Dept. of Microbiology, University of California, Los Angeles

Michael Botchan, Dept. of Molecular Biology, University of California, Berkeley

Pierre Chambon, Laboratoire de Génétique Moléculaire des Eucaryotes du CNRS, Institut de Chimie Biologique, Strasbourg, France

Tau Rong Chiang, Schering Corporation, Bloomfield, New Jersey

Bryan Cullen, Hoffmann-La Roche Inc., Nutley, New Jersey

Donald DeFranco, Dept. of Biochemistry, University of California, San Francisco

Jean de Villiers, Genetics Institute, Boston, Massachusetts

Anne Ephrussi, Cancer Center, Massachusetts Institute of Technology, Cambridge

S. Carl Falco, E.I. du Pont de Nemours & Co., Wilmington, Delaware

Michael Fried, ICRF, Lincoln's Inn Fields, London, England

Yakov Gluzman, Cold Spring Harbor Laboratory, New York

Cori Gorman, Dept. of Molecular Biology, Genentech Institute, San Francisco, California

Peter Gruss, Institut für Mikrobiologie, University of Heidelberg, Federal Republic of Germany

Leonard Guarente, Dept. of Biology, Massachusetts Institute of Technology, Cambridge

Dennis Harris, Amersham International plc, Buckinghamshire, England

John A. Hassell, Dept. of Microbiology and Immunology, University of Montreal, Canada

Patrick Hearing, Dept. of Microbiology, State University of New York, Stony Brook

Nathaniel Heintz, Laboratory of Biochemistry and Molecular Biology, Rockefeller University, New York, New York

Winship Herr, Cold Spring Harbor Laboratory, New York

James Hicks, Cold Spring Harbor Laboratory, New York

Peter Johnson, Dept. of Embryology, Carnegie Institution of Washington, Baltimore, Maryland

Katherine Jones, Dept. of Biochemistry, University of California, Berkeley

George Khoury, Laboratory of Molecular Virology, National Institutes of Health, Bethesda, Maryland

Sheila P. Little, Lilly Research Laboratories, Indianapolis, Indiana

Steve McKnight, Dept. of Embryology, Carnegie Institution of Washington, Baltimore, Maryland

Thomas Maniatis, Dept. of Biochemistry and Molecular Biology, Harvard University, Cambridge, Massachusetts

Michele Manos, Cetus Corporation, Emeryville, California

Don Merlo, Agrigenetics Advanced Research Division, Madison, Wisconsin

Eliot Meyerowitz, Division of Biology, California Institute of Technology, Pasadena

Kim Nasmyth, Medical Research Council, Laboratory of Molecular Biology, Cambridge, England

Sharon Ogden, Monsanto Company, Chesterfield, Missouri

Richard Palmiter, Howard Hughes Medical Institute, Research Laboratories, Seattle, Washington

Leonard E. Post, Upjohn Company, Kalamazoo, Michigan

Mark Ptashne, The Biological Laboratories, Harvard University, Cambridge, Massachusetts

Walter Schaffner, Institut für Molekularbiologie II der Universitat Zurich, Switzerland

Ranjan Sen, Whitehead Institute for Biomedical Research, Cambridge, Massachusetts

Seshi Takahasi, Mitsui Toatsu Chemicals, Inc., New York, New York

Richard Treisman, Medical Research Council, Laboratory of Molecular Biology, Cambridge, England

Michael D. Walker, University of California, San Francisco

Pieter Wensink, Rosenstiel Center, Brandeis University, Waltham, Massachusetts

Alan G. Wildeman, CNRS, Institute de Chimie Biologique, Strasbourg, France

Moshe Yaniv, Dept. of Molecular Biology, Institut Pasteur, Paris, France

Edward B. Ziff, Dept. of Biochemistry, New York University Medical School, New York

Kai Zinn, Dept. of Biochemistry and Molecular Biology, Harvard University, Cambridge, Massachusetts

Preface

Regulation of transcription, particularly its initiation step, is an important phase in the programmed expression of genes in eukaryotic organisms. Two years ago the first conference at Cold Spring Harbor on the topic of the initiation of transcription took place, with particular emphasis on the structure of enhancers and regulated promoters. The two years since then have been especially fruitful in the collecting of data regarding the *cis-* and *trans-*acting elements involved in initiation. The DNA sequences responsible for enhancement activities have now been mapped much more precisely. In addition, protein interactions with these DNA sequences have been identified both in vitro and in vivo by the use of footprinting techniques. The importance of the initiation of transcription as a step in the tissue-specific expression of a great variety of genes has also been clearly documented. Furthermore, these experiments have been taken to their logical conclusion by demonstrating tissue-specific expression in transgenic organisms.

These exciting results were presented during the second conference on eukaryotic transcription, which was held at the Banbury Center of Cold Spring Harbor Laboratory on March 24–27, 1985. *Eukaryotic Transcription* summarizes these talks in the form of extended abstracts provided by the speakers, who discussed various experimental systems and a broad spectrum of genes, whose origins ranged from viral to human.

I am grateful to James D. Watson, Director of Cold Spring Harbor Laboratory, for making available the facilities of the Banbury Center, as well as Mike Shodell and his staff of the Banbury Center and Nancy Ford and Doug Owen of the Publications Department.

Yakov Gluzman
June 1985

The meeting on The Role of *cis*- and *trans*-acting Elements in the Initiation of Eukaryotic Transcription was funded entirely by proceeds from the Laboratory's Corporate Sponsor Program, whose members provide core support for Cold Spring Harbor and Banbury meetings:

Agrigenetics Corporation
American Cyanamid Company
Amersham International plc
Becton Dickinson & Company
Biogen S.A.
Cetus Corporation
Ciba-Geigy Corporation
CPC International, Inc.
E.I. du Pont de Nemours & Company
Genentech, Inc.
Genetics Institute
Hoffmann-La Roche Inc.
Johnson & Johnson
Eli Lilly and Company
Mitsui Toatsu Chemicals, Inc.
Monsanto Company
Pall Corporation
Pfizer Inc.
Schering-Plough Corporation
Smith Kline & French Laboratories
Upjohn Company

Contents

Introduction

W. Schaffner

Institut für Molekularbiologie II der Universität Zürich
Hönggerberg CH-8093 Zürich, Switzerland

There are two distinct levels of control in the initiation of transcription by RNA polymerase: the frequency of initiation and the site of initiation. In eubacteria the situation appears to be straightforward. Conserved DNA sequences that bind RNA polymerase are located at about 10 bp and 35 bp upstream (5') of the initiation point (Fig. 1) (Rosenberg and Court 1979; Siebenlist et al. 1980). In addition to these constitutive promoter elements, there are sequences governing negative and/or positive regulation. For negative regulation a repressor protein binds to operator DNA, which often overlaps with the RNA polymerase binding site. If a repressor is bound to the operator, transcription is inhibited. On the other hand, some promoters are subject to positive regulation whereby an activator protein binds to a specific DNA sequence in the upstream promoter region and, presumably by making contact with RNA polymerase, stimulates transcription (Hochschild et al. 1983).

In higher eukaryotes, detailed information on the regulation of transcription could not be obtained until the development of the techniques of "surrogate" or "reverse genetics" (i.e., in vitro manipulation of cloned genes followed by functional testing). Initially the situation seemed very similar to that in bacteria. A conserved AT-rich DNA sequence motif was found to be about 30 bp upstream of most eukaryotic transcription units, and by in vitro transcription assays this so-called TATA box was shown to be an essential promoter component (Fig. 1) (Chambon et al. 1984). An additional conserved sequence of consensus GG^C_TCAATCT, the so-called CAAT box, was found upstream of the TATA box in virtually all globin genes and in several other genes (Benoist et al. 1980; Efstratiadis et al. 1980; McKnight et al., this volume). The search for additional promoter components by sequence analysis proved difficult, however, because of the multitude of sequences located further upstream of the TATA box.

A puzzling observation in the study of eukaryotic gene expression was made by Grosschedl and Birnstiel (1980) during the injection of cloned sea urchin histone genes into *Xenopus* oocytes. They found

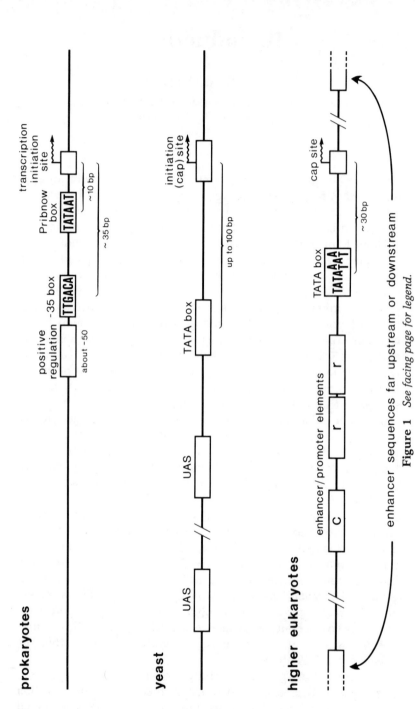

Figure 1 *See facing page for legend.*

Figure 1 (*see facing page*) Simplified model of transcription control. In *prokaryotic* (eubacterial) promoters, the −35 box and the Pribnow box interact with RNA polymerase. Positive regulation can be mediated by a neighboring upstream element binding an activator protein, which in turn facilitates transcription by making contact with RNA polymerase. Spacing of these control elements is quite strict. Prokaryotic promoters are often negatively regulated by a repressor protein binding to an operator sequence near the transcription initiation site, but no such operator is indicated.

In *yeast*, sequences near or at the transcription start site determine the initiation point, whereas the TATA box is located at variable distances and is mainly required for efficient transcription. An upstream activation site (UAS) may bind one or more factors that regulate transcription, and there may be several UASs for regulation by more than one stimulus. UASs can be located far upstream and in either orientation but apparently not behind the transcription initiation point.

Higher eukaryotes, with mammals as the typical example, show the greatest freedom in the position of transcription control elements with respect to the initiation site. Whereas the TATA box at a position around −30 determines the cap site, other elements can be located in the immediate upstream region and also far upstream or downstream of the cap site, where they may control additional functions such as DNA replication.

Regulated gene expression involves a modular arrangement of both constitutive (c) and regulatory (r) DNA elements and the corresponding protein factors (see also Serfling et al., this volume). Upon stimulation, a regulatory factor, perhaps after an allosteric conformation change, recognizes the corresponding DNA motif and thus may form a bridge between constitutive factors (the latter including the TATA-box-binding protein). The number and spatial arrangement of these motifs presumably determines the basal vs. induced level of transcription; e.g., maximal transcription can be limited by the number of enhancer-promoter elements ("the longer the better").

Exploiting remote control by the enhancer effect, eukaryotes seem to realize a great variety of regulatory networks by modular combinations of a set of DNA elements.

3

that sequences located more than 100 bp upstream of the initiation (cap) site of a histone H2A gene could act in either orientation to restore efficient transcription (Grosschedl and Birnstiel 1980). This orientation independence of upstream promoter elements, even though they often do not show dyad symmetry, has since been found to be a property of most promoter elements of eukaryotes from yeast to man. In parallel with these studies came the discovery of the enhancer effect in our lab (Banerji et al. 1981) as well as in the labs of Chambon (Moreau et al. 1981) and Berg (Fromm and Berg 1983). For example, sequences upstream of the "early" promoter of the simian virus 40 (SV40) were found to stimulate transcription of a linked β-globin gene by more than two orders of magnitude in either orientation and over distances of more than 3000 bp, even from a position downstream of the gene (Banerji et al. 1981). Such an enhancer effect had not been previously described for any promoter component. These sequences, referred to as the enhancer, were narrowed to an upstream promoter region that was known to be required for efficient early gene transcription (Benoist and Chambon 1981; Gruss et al. 1981). In the following years, in many unrelated viruses, enhancers were discovered (for reviews, see Gluzman and Shenk 1983; Khoury and Gruss 1983; Yaniv 1984; Picard 1985; Serfling et al. 1985) that, in experimental constructions, fulfilled the following strict criteria:

1. strong activation of transcription of a linked gene from the correct initiation (cap) site;
2. activation of transcription independent of orientation;
3. ability to function over long distances of more than 1000 bp whether from an upstream or downstream position relative to the cap site; and
4. in the cases tested, preferential stimulation of transcription from the most proximal of two tandem promoters (de Villiers et al. 1983; Wasylyk et al. 1983; T. Kadesch and P. Berg, pers. comm.).

The first cellular enhancer was discovered in the immunoglobulin (Ig) heavy-chain gene (Banerji et al. 1983; Gillies et al. 1983; Neuberger 1983). This was interesting for three reasons:

1. It showed that the enhancer effect was not a viral peculiarity.
2. The Ig enhancer stimulated transcription in a cell-type-specific manner, being the first genetic element described to have such a property.
3. The Ig enhancer was found to be located within the gene, in a position downstream of the cap site, thus realizing in a natural

situation the potential of enhancers shown earlier by in vitro constructions (Banerji et al. 1981).

Although DNA sequences with enhancer activity are most often found upstream of the transcription unit between bp −100 and −300, they are also found at other positions, such as within an intron for the Ig heavy-chain enhancer or between two genes. With the analysis of more enhancers, we have come to realize that upstream promoter components and enhancers overlap physically and functionally, and that the enhancer is perhaps not so much a physical entity with clearly definable boundaries, but rather that there is an "enhancer effect." The enhancer effect can be exerted by combinations of a variety of DNA sequence motifs. Although there are sequence motifs that seem to occur preferentially in one orientation and just upstream of the TATA box, like the CAAT motif, it now appears that there is no strict distinction between "promoter"- and "enhancer"-specific components. For example, upstream promoter elements of the inducible metallothionein-I gene show typical enhancer behavior when detached from the TATA box and linked to a test gene (Serfling et al., this volume). Therefore, these upstream elements can be viewed, depending on their spatial context, to be part of either a promoter region or an enhancer. Nevertheless, for convenience, I shall continue to call a DNA segment with enhancer activity an enhancer.

Viral Enhancers for Early Gene Expression, DNA Replication, and Oncogenesis

Among the many known enhancers, the one of SV40 is the best characterized and, therefore, will be presented here as the prototypical enhancer. The genome of SV40, a member of the papovaviruses, is a short, circular DNA molecule of 5.2 kb (for review, see Tooze 1981). It contains two transcription units: The early transcription unit is expressed during the initial phase of viral infection, while the late transcription unit is expressed at the onset of viral DNA replication. The transcription units are transcribed in opposite directions, and their promoters overlap each other near the origin of replication. Essential elements for the expression of the early genes are the two 72-bp repeats located within the promoter/regulatory region of the early and late genes. These 72-bp repeats turned out to be the major contributors to the enhancer effect.

The SV40 enhancer has been shown to act in a wide variety of tissues and hosts. In addition to its activity in mammalian cells, it also stimulates transcription when introduced into amphibian cells

5

(our unpubl. observations) and even algal cells (Neuhaus et al. 1984). In contrast, the enhancer of the mouse polyoma virus, a papovavirus like SV40, shows a distinct host-cell preference. Whereas the SV40 enhancer is about equally active in mouse and primate cells, the polyoma enhancer is approximately four times as active in mouse cells than in primate cells (de Villiers et al. 1984).

Apart from a preference for cells of the host species, certain enhancers, such as the one of polyoma virus, also show a pronounced cell-type specificity (Herbomel et al. 1984; see also Gorman et al., this volume). Especially striking is the enhancer of JC virus (JCV), the papovavirus agent of the demyelinating disease, progressive multifocal leukoencephalopathy (Kenney et al. 1984). In fetal glial cells, the 98-bp-long enhancer repeat of JCV is more than 20 times as strong as in HeLa cells. It shares a discontinuous stretch of about 20 bp with the so-called identifier (ID) elements, a class of middle repetitive sequences of the human genome that may be involved in the regulation of brain-specific gene expression (Sutcliffe et al. 1984).

Enhancer elements have also been detected in other papovaviruses, hepatitis B virus, several retroviruses, the early region of adenoviruses, and the immediate-early regions of herpes viruses (for review, see Gluzman and Shenk 1983; Khoury and Gruss 1983; Yaniv 1984; Picard 1985; Serfling et al. 1985; see also Hearing and Shenk; Fried et al.; both this volume). In retroviruses, enhancers play a crucial role in oncogenesis (Payne et al. 1982; Cullen et al. 1985; Weber and Schaffner 1985b) and are, at least in some cases, responsible for the tissue tropism of cancer formation (Celander and Haseltine 1984; Stewart et al. 1984). Whereas viral enhancers are required for immediate-early gene expression and, at least in some cases, for viral DNA replication (de Villiers et al. 1984; Hassell et al., this volume), late-early or late gene expression is often stimulated by so-called *trans*-activating proteins, such as adenovirus E1A, which seem to act in an indirect manner by mobilizing certain transcription factors (Yoshinaga et al., this volume; see also Green et al. 1983; Robbins and Botchan; Brady et al.; both this volume).

Tissue-specific Ig Enhancers

Many principles of viral gene expression are also used in the expression of cellular genes, and thus it was expected that enhancers would be regulatory elements of cellular genes, as well (Banerji et al. 1981). This question was approached by looking for an enhancer in the Ig heavy-chain gene. Ig genes consist of several gene seg-

6

ments. In the course of lymphoid cell differentiation, the Ig heavy-chain genes are assembled by the fusion of one of more than one hundred variable (V) gene segments to a diversity (D) and a joining (J) segment that, in itself, is separated by a large intron from a constant (C) segment. Until recently it was not known how the promoter of the rearranged V segment could be selectively transcribed while the many unrearranged V promoters are silent in differentiated B cells. As mentioned above, a strong enhancer is situated in the intron between the J and C segments of the mouse Cμ gene (Banerji et al. 1983; Gillies et al. 1983; Neuberger 1983). The enhancer, therefore, is located 3' to the promoter and within the Ig transcription unit. The Ig heavy-chain enhancer has the same characteristics as a typical viral enhancer: It works in both orientations and acts over long distances on heterologous promoters. In addition it shows a pronounced tissue specificity functioning only in lymphoid cells. The Ig heavy-chain enhancer was the first genetic element identified to confer a cell-type specificity (Banerji et al. 1983; Gillies et al. 1983; Neuberger 1983).

The existence of a transcription enhancer within the J-C intron seemed to clarify two basic processes in B-cell differentiation. First is the already mentioned observation that the promoter of a rearranged V segment is selectively activated. This segment, upon rearrangement, is brought into the proximity of the Ig enhancer and subsequently becomes transcriptionally active. Moreover, the heavy-chain enhancer is situated in front of the switch region and it remains, therefore, unchanged after the C-region class switch, an event in which the "early" Cμ-Cδ segments are replaced by the "late" Cγ, Cϵ, and Cα segments. In light of the latter observation, it was surprising to learn that spontaneous deletions of the heavy-chain enhancer that occurred in IgM- and IgD-producing myeloma cells did not affect Ig expression (Klein et al. 1984; Wabl and Burrows 1984). From our recent experiments, we suggest that the Ig enhancer is required for establishing stable transcriptional complexes early in B-cell differentiation but is dispensable for the maintenance of Ig gene transcription (S. Klein, T. Gerster, D. Picard, A. Radbruch, and W. Schaffner, unpubl.). An enhancer element has been identified in the Ig $\varkappa$ light-chain gene (Picard and Schaffner 1984; Queen and Stafford 1984). It resembles the heavy-chain enhancer in its location and tissue specificity. In transfection assays, it is, however, less active than the heavy-chain enhancer (Picard and Schaffner 1984). The $\varkappa$ enhancer sequence is evolutionarily conserved among different mammals, and like the chromatin of the SV40 and polyoma enhancers (see below), the chromatin region of this enhancer is preferentially accessible to DNase I and restriction enzymes (Parslow and Granner 1982).

In addition to the cell-type specificity of the enhancers, the Ig promoters also show a pronounced cell-type preference for lymphoid cells (Falkner et al. 1984; Foster et al. 1985; Mason et al. 1985).

Does Every Cellular Gene Have an Enhancer?

Unlike the Ig genes, the overwhelming majority of cellular genes are not rearranged during cell differentiation. Therefore, it was of interest to see whether enhancers are also regulatory elements of other cellular genes and, if so, where they are located – within the transcription unit, as the Ig enhancer, or immediately upstream of the gene, as for the overwhelming majority of viral enhancers. More recently, it has been found that insulin gene expression is mediated by tissue-specific enhancer sequences located about 100–400 bp upstream of the transcription initiation site (Walker et al. 1983; Edlund et al., this volume). A transforming gene (T antigen) under the control of the insulin enhancer-promoter was injected into mouse eggs. The resulting transgenic mice showed T-antigen expression that was confined to the β cells of the pancreatic Langerhans islets. These mice also had a high incidence of β-cell tumors, thus establishing a new approach to targeted oncogenesis (Hanahan 1985; see also Stewart et al. 1984). Tissue-specific enhancer elements have also been found upstream of rat chymotrypsin (Walker et al. 1983; Edlund et al., this volume) and albumin genes (Ott et al. 1984) and of the Eβ gene of the mouse major histocompatibility gene family (Gillies et al. 1984). Inducible enhancers, first identified in mouse mammary tumor virus (Chandler et al. 1983), have since been found in cellular genes as well, such as the metallothionein gene (Serfling et al., this volume) the β-interferon gene (Goodbourn et al., this volume), and the *fos* proto-oncogene (Treisman; see also Greenberg et al.; both this volume). Recent findings with several genes suggest that enhancers can be located at various positions within developmentally regulated DNA domains. There they may also contribute to the control of additional functions such as DNA replication, as recently shown in viral systems (de Villiers et al. 1984; Hassell, this volume; R. Krumlauf; A. Sippel; both pers. comm.).

The analyses of chromosomal enhancers demonstrate that viral and chromosomal enhancers belong to the same type of transcriptional control elements. Viral as well as cellular enhancers stimulate transcription from heterologous promoters over long distances in an orientation-independent manner, and both types of enhancers can contain DNA sequences conferring tissue-specific gene expression. Although it is not known at present whether enhancer elements are essential for the activation of all cellular genes, it is likely that the difference in structure between strongly and weakly transcribed promoters is of a quantitative rather than a qualitative nature. That is, the many sequence motifs used to build an enhancer of a strongly expressed gene could be present in only a few copies in the promoter region of a typical "housekeeping" gene (see Serfling et al.; McKnight et al.; Heintz et al.; all this volume).

Relation of Enhancer Structure to Activity: The Longer the Better

Direct repeats are often found within enhancers. However, in contrast to a widespread belief, the 72-bp direct repeat of SV40 is not *the* enhancer. There is a considerable contribution from nonrepeated sequences further upstream with respect to the early initiation sites (Weber et al. 1984; Herr and Gluzman 1985; Herr et al.; Wildeman et al.; Cereghini et al.; all this volume). Direct tandem repeats, usually 50–130 bp long, are also found within the enhancers of BK virus (Rosenthal et al. 1983), Moloney sarcoma virus (Levinson et al. 1982), and mouse cytomegalovirus (Dorsch-Häsler et al. 1985). Other enhancers, however, do not show obvious repeats, such as the ones of polyoma virus strain A2 (de Villiers and Schaffner 1981), hepatitis B virus (Shaul et al. 1985; A. Tognoni et al., unpubl.), Rous sarcoma virus (Luciw et al. 1983; Laimins et al. 1984; Cullen et al. 1985; Weber and Schaffner 1985b), and the majority of cellular enhancers. An SV40 enhancer lacking one of the two 72-bp repeats is sufficiently active to support viral growth (Gruss et al. 1981; Herr and Gluzman 1985). Sequence repetition per se is not an essential feature of enhancers, but it obviously potentiates activity. For example, polyoma variants selected for growth in cell types where the wild-type strain does not replicate often are found to have duplicated a specific enhancer segment (Sekikawa and Levine 1981; Vasseur et al. 1982). The activity of a severely debilitated enhancer can be resurrected by duplication in vivo under selective pressure (Swimmer and Shenk 1984; Weber et al. 1984; Herr and Gluzman 1985; Herr et al., this volume). Even more striking are the cases in which dimerization of an arbitrarily truncated enhancer was performed in vitro and subsequent functional testing demonstrated significant activity (de Villiers et al. 1984; Laimins et al. 1984; Weber et al. 1984). Enhancers seem to be generally forgiving of all kinds of sequence manipulations. Human cytomegalovirus contains a very long, extremely strong enhancer that can functionally replace the enhancer of SV40 virus (Boshart et al. 1985). A number of extended deletions of the cytomegalovirus enhancer were constructed and tested for activity. Surprisingly, none had lost the ability to support growth of SV40 virus. From these data it became clear that various subsegments of this long enhancer can individually substitute for the SV40 enhancer.

The Modular Units of Enhancer Activity

Closer inspection of the cytomegalovirus enhancer has revealed the presence of four different families of repeated sequence motifs of 17 bp, 18 bp, 19 bp, and 21 bp in length scattered throughout the approximately 450-bp enhancer sequence. Among these motifs, the 18-bp and 19-bp repeats are most remarkable. The 18-bp, imperfectly repeated motif of consensus ^A_CCTAACGGGACTTTCCAA (Boshart et al. 1985) harbors a sequence referred to as the enhancer "core," which is common to several other enhancers (usually read in the opposite orientaton as GTGG$^{AAA}_{TTT}$G) (Weiher et al. 1983). The 19-bp repeat of consensus sequence C̲C̲C̲C̲ATTGACGTCAAT̲G̲G̲G̲ is organized as an inverted repeat and is present in the cytomegalovirus enhancers of human, monkey, and mouse origin. From these and other data (Boshart et al. 1985; Dorsch-Häsler et al. 1985; Herr and Gluzman 1985; Cereghini et al; DeFranco et al.; Ephrussi et al.; Hassell; Herr et al.; Wildeman et al.; all this volume), we conclude that enhancers generally have a modular structure with highly redundant information in which conserved motifs of several kinds are separated by less-conserved DNA sequences (Fig. 1). (A modular structure has also been suggested for the immediate upstream elements of the β-globin and thymidine kinase promoters by Weissmann and colleagues [Cochran and Weissmann 1984].) Presumably these enhancer-promoter elements bind cellular transcription factors, either of a constitutive (such as the Sp1 factor binding to the CCGCCC motif; Gidoni et al. 1984; Jones and Tjian, this volume) or a regulatory type (such as a steroid hormone receptor; Scheidereit et al. 1983; DeFranco et al.; Gaub et al.; both this volume) or the factor for metal induction of metallothionein genes (Seguin et al. 1984; Palmiter et al.; Serfling et al.; both this volume). Several of these sequence motifs have been shown by means of "footprinting" and "gel retardation" assays to bind specific protein factors (Cereghini et al.; Ephrussi et al.; Jones and Tjian; Sen and Baltimore; Wildeman et al.; all this volume), and it can be expected that the isolation and analysis of these factors will further our understanding of the mechanisms of gene regulation (see also below). It appears that the multitude of constitutive sequence motifs can functionally replace each other since enhancers can be constructed in which any given type of consensus sequence is missing (Weber et al. 1984; Boshart et al. 1985; Herr and Gluzman 1985; Herr and Gluzman; Hassell; both this volume). A model of the interaction of constitutive and regulatory components by "bridging" of transcription factors is presented in Figure 1 (see also McKnight et al.; Serfling et al.; both this volume).

The Same Principles Everywhere?

So far, no examples of remote control of gene transcription have been reported for eubacteria. However, similar principles seem to govern the control of transcription of protein-coding genes in all eukaryotes, including yeast. In *Drosophila*, the available information suggests that tissue-specific enhancers control gene expression during development and differentiation in a manner very similar to, if not identical with, that in mammalian systems (Meyerowitz et al.; Posakony et al.; Wensink et al.; all this volume).

Among all nonmammalian eukaryotes, the most detailed information on promoter structure has been obtained from yeast, where some notable differences with higher eukaryotes have become apparent. First, the TATA box is found at a variable distance up to 100 bp from the site of transcription initiation. This TATA box determines the level of transcription, but where transcription initiates is encoded by the intiation site itself, presumably by some "initiation box" sequence (e.g., see Guarente, this volume). Regulated yeast promoters, similar to mammalian enhancer-promoter regions, also show a modular structure that contains, in addition to the initiation box and a TATA box, one or more of the so-called upstream activation sites (UASs). These UASs are recognized by specific regulatory proteins in response to gene-specific phsyiological signals (Guarente and Hoar 1984; Struhl 1984; Guarente; Nasmyth et al.; Ptashne et al.; all this volume). The UASs can exert their effect in either orientation and over large distances of as much as 1500 bp (Struhl 1984; Nasmyth et al.; Ptashne et al.; both this volume), but unlike mammalian enhancers they apparently cannot activate transcription from a downstream position (Guarente and Hoar 1984; Struhl 1984). It remains to be seen whether some features of promoter structure are shared between genes of yeast on one hand and those coding for small, nuclear URNAs transcribed by polymerase II and ribosomal RNA genes (transcribed by RNA polymerase I) in higher eukaryotes on the other. In the latter cases, upstream enhancer-like sequences have been described that work at a distance and in either orientation (Reeder 1984; Mattaj et al. 1985) but apparently not from a downstream position. Nevertheless, there are *cis*-acting control sequences in yeast with an effect from a downstream position. So-called silencers in the mating-type locus exert a negative control, in that a transcription unit linked to such silencer sequences is efficiently repressed (Brand et al. 1985).

Facts and Speculations about the Enhancer Mechanism

How does an enhancer enhance transcription? First of all, it functions in *cis*, i.e., only on the same DNA molecule (Banerji et al. 1981; Moreau et al. 1981). It has been shown that an active enhancer ensures a high density of RNA polymerase II molecules over a linked gene, resulting in a high rate of transcription (Treisman and Maniatis 1985; Weber and Schaffner 1985a). Also, there is a peculiar chromatin structure associated with enhancers: The enhancer region of both SV40 and polyoma virus is a nucleosome-free region (Saragosti et al. 1980; Herbomel et al. 1981; Jongstra et al. 1984). From subsequent work it became evident that this region, which is also hypersensitive to nucleases, is associated with a variety of transcription factors (Davison et al. 1983; Gidoni et al. 1984; Schöler and Gruss 1984; Cereghini et al.; Jones and Tjian; Wildeman et al.; all this volume; see also model presented in Fig. 1 and in Serfling et al., this volume). It was speculated that the "open window" of such a nucleosome-free gap is an entry site for RNA polymerase (Saragosti et al. 1980; Herbomel et al. 1981; Moreau et al. 1981; Allan et al. 1984; Jongstra et al. 1984). An enhancer could also facilitate, in a large segment of DNA, binding of chromatin proteins specific for active transcription units. This would result in the formation of stable transcription complexes analogous to the ones described for 5S rRNA genes by Brown and his colleagues (Bogenhagen et al. 1982; see also Reeder 1984; Ryoji and Worcel 1984; Mattaj et al. 1985). One can even speculate that in certain cases an enhancer sequence which was required for the organization of an active chromatin domain becomes dispensable for the maintenance of stable transcription complexes. This would explain why the Ig heavy-chain enhancer can be deleted in B lymphocytes without affecting the high level of transcription (see above). In other cases, however, like the inducible enhancer of the metallothionein gene, activation must be reversible.

The enhancer effect is most evident in a living cell, but some effect is also seen in cell-free transcription extracts. In vitro, it has been observed that an SV40 enhancer stimulates transcription up to 10-fold from a position close to the transcription initiation site

(Sassone-Corsi et al. 1984; Sergeant et al. 1984; Wildeman et al. 1984; see also Sen and Baltimore; Schöler et al.; both this volume). With the SV40 enhancer downstream of the rabbit β-globin gene, however, there is even less transcription than without an enhancer (Sergeant et al. 1984; see also Sen and Baltimore, this volume). (In contrast, the SV40 enhancer, whether located upstream or downstream of the β-globin gene, stimulates in vivo transcription by more than two orders of magnitude [Banerji et al. 1981; Treisman et al. 1983; Sergeant et al. 1984].) Such in vitro transcription systems will undoubtedly prove useful for the analysis of the factors involved in enhancer action. Nevertheless, it is at present a matter of debate whether the full enhancer effect can be achieved in vitro with soluble transcription extracts or whether some higher-order nuclear structures are also required. We tend to believe the latter, i.e., that an attachment to a nuclear compartment is required for a full manifestation of the effect. Such an attachment effect can be transmitted over large distances by "folding back" or "reeling through" long DNA segments and/or by changes in the linking number of an entire DNA domain.

CONCLUSIONS

Enhancers were originally identified as long-range activators of gene transcription in higher eukaryotes and, in addition, they were the first DNA sequences found to confer tissue specificity. These properties had set them apart from previously described upstream promoter elements of eukaryotic genes. Intense investigation over the past two years of the sequences involved in eukaryotic gene regulation has suggested that enhancer and promoter elements overlap both physically and functionally, being distinguishable quantitatively but not qualitatively. Our current view is that enhancers and upstream regulatory elements are composed of a modular arrangement of short sequence motifs, each with a specific function in conferring inducibility, tissue specificity, or a general enhancement of transcription. These motifs are presumably binding sites for nuclear proteins whose mechanism of action remains to be elucidated.

The control of transcription in higher eukaryotes appears more complex, and at first sight, less economical than in prokaryotes. However, the evolution of regulatory sequences to contain a variety of sequence motifs that respond to different stimuli and that are flexible with regard to orientation and distance from the transcription initiation site may have been crucial for the establishment of regulatory networks. Such networks must have been of paramount importance for the evolution of multicellular organisms.

ACKNOWLEDGMENTS

I am indebted to Maria Jasin, Edgar Serfling, Fritz Ochsenbein, and Benno Müller-Baden for help and valuable suggestions, and to Max Birnstiel, Greg Gilmartin, and Fred Schaufele for critical reading of the manuscript. I am also grateful to Yasha Gluzman and Doug Owen for their encouragement and patience.

Work done in our lab was supported by the Schweizerischer Nationalfonds and the Kanton Zürich.

REFERENCES

Allan, M., J.-D. Zhu, P. Montague, and J. Paul. 1984. Differential response of multiple ε-globin cap sites to cis- and trans-acting controls. *Cell* **38:** 399.

Banerji, J., L. Olson, and W. Schaffner. 1983. A lymphocyte-specific cellular enhancer is located downstream of the joining region in immunoglobulin heavy chain genes. *Cell* **33:** 729.

Banerji, J., S. Rusconi, and W. Schaffner. 1981. Expression of a globin gene is enhanced by remote SV40 DNA sequences. *Cell* **27:** 299.

Benoist, C. and P. Chambon. 1981. *In vivo* sequence requirements of the SV40 early promoter region. *Nature* **290:** 304.

Benoist, C., K. O'Hare, R. Breathnach, and P. Chambon. 1980. The ovalbumin gene-sequence of putative control regions. *Nucleic Acids Res.* **8:** 127.

Bogenhagen, D.F., W.M. Wormington, and D.D. Brown. 1982. Stable transcription complexes of *Xenopus* 5S RNA genes: A means to maintain the differentiated state. *Cell* **28:** 413.

Boshart, M., F. Weber, G. Jahn, K. Dorsch-Häsler, B. Fleckenstein, and W. Schaffner. 1985. A very strong enhancer is located upstream of an immediate early gene of human cytomegalovirus. *Cell* **41:** 521.

Brand, A.H., L. Breeden, J. Abraham, R. Sternglanz, and K. Nasmyth. 1985. Characterization of a "silencer" in yeast: A DNA sequence with properties opposite to those of a transcriptional enhancer. *Cell* **41:** 41.

Celander, D. and W.A. Haseltine. 1984. Tissue-specific transcription preference as a determinant of cell tropism and leukaemogenic potential of murine retroviruses. *Nature* **312:** 159.

Chambon, P., A. Dierich, M.-P. Gaub, S. Jakowlew, J. Jongstra, A. Kurst, J.-P. LePennec, P. Oudet, and T. Reudelhuber. 1984. Promoter elements of genes coding for proteins and modulation of transcription by estrogens and progesterone. *Recent Prog. Horm. Res.* **40:** 1.

Chandler, V.L., B.A. Maier, and K.R. Yamamoto. 1983. DNA sequences bound specifically by glucocorticoid receptor in vitro render a heterologous promoter hormone responsive in vivo. *Cell* **33:** 489.

Cochran, M.D. and C. Weissmann. 1984. Modular structure of the β-globin and the TK promoters. *EMBO J.* **3:** 2453.

Cullen, B.R., K. Raymond, and G. Ju. 1985. Functional analysis of the transcription control region located within the avian retroviral long terminal repeat. *Mol. Cell. Biol.* **53:** 438.

Davison, B.L., J.-M. Egly, E.R. Mulvihill, and P. Chambon. 1983. Formation of stable preinitiation complexes between eukaryotic class B transcription factors and promoter sequences. *Nature* **301:** 680.

de Villiers, J. and W. Schaffner. 1981. A small segment of polyoma virus DNA enhances the expression of a cloned rabbit β-globin gene over a distance of at least 1400 base pairs. *Nucleic Acids Res.* **9:** 6251.

de Villiers, J., L. Olson, J. Banerji, and W. Schaffner. 1983. Analysis of the transcriptional enhancer effect. *Cold Spring Harbor Symp. Quant. Biol.* **47:** 911.

de Villiers, J., L. Olson, C. Tyndall, and W. Schaffner. 1982. Transcriptional "enhancers" from SV40 and polyoma virus show a cell type preference. *Nucleic Acids Res.* **10:** 7965.

de Villiers, J., W. Schaffner, C. Tyndall, S. Lupton, and R. Kamen. 1984. Polyoma virus DNA replication requires an enhancer. *Nature* **312:** 242.

Dorsch-Häsler, K., G.M. Keil, F. Weber, W. Schaffner, and U.H. Koszinowski. 1985. A long and complex enhancer activates transcription of the gene coding for the highly abundant immediate early mRNA in murine cytomegalovirus. *Proc. Natl. Acad. Sci.* (in press).

Efstratiadis, A., J.W. Posakony, T. Maniatis, R.M. Lawn, C. O'Connell, R.A. Spritz, J.K. Riel, B.G. de Forget, S.W. Weissmann, J.L. Slightom, A.E. Blechl, O. Smithies, F.E. Baralle, C.C. Shoulders, and N.J. Proudfoot. 1980. The structure and evolution of the human β-globin gene family. *Cell* **21:** 653.

Falkner, F.G., E. Neumann, and H.G. Zachau. 1984. Tissue specificity of the initiation of immunoglobulin kappa gene transcription. *Hoppe-Seyler's Z. Physiol. Chem.* **365:** 1331.

Foster, J., J. Stafford, and C. Queen. 1985. An immunoglobulin promoter displays cell-type specificity independently of the enhancer. *Nature* **315:** 423.

Fromm, M. and P. Berg. 1983. Simian virus 40 early- and late-region promoter functions are enhanced by the 72 base pair repeat inserted at distant locations and inverted orientations. *Mol. Cell. Biol.* **3:** 991.

Gidoni, D., W.S. Dynan, and R. Tjian. 1984. Multiple contacts between a mammalian transcription factor and its cognate promoters. *Nature* **312:** 409.

Gillies, S.D., V. Folsom, and S. Tonegawa. 1984. Cell-type-specific enhancer element associated with a mouse MHC E_β. *Nature* **310:** 594.

Gillies, S.D., S.L. Morrison, V.T. Oi, and S. Tonegawa. 1983. A tissue-specific transcription enhancer element is located in the major intron of a rearranged immunoglobulin heavy chain gene. *Cell* **33:** 717.

Gluzman, Y. and T. Shenk, eds. 1983. *Current communications in molecular biology: Eukaryotic gene expression.* Cold Spring Harbor Laboratory, Cold Spring Harbor, New York.

Green, M.R., R. Treisman, and T. Maniatis. 1983. Transcriptional activation of cloned human β-globin genes by viral immediate-early gene products. *Cell* **35:** 137.

Grosschedl, R. and Birnstiel, M. 1980. Spacer DNA sequences upstream of the TATAAATA sequence are essential for the promotion of H2A histone gene transcription *in vivo*. *Proc. Natl. Acad. Sci.* **77:** 7102.

Gruss, P., R. Dhar, and G. Khoury. 1981. Simian virus 40 tandem repeated sequences as an element of the early promoter. *Proc. Natl. Acad. Sci.* **78:** 943.

Guarente, L. and E. Hoar. 1984. Upstream activation sites of the *CYC1* gene of *Saccharomyces cerevisiae* are active when inverted but not when placed downsteam of the "TATA box." *Proc. Natl. Acad. Sci.* **81:** 7860.

Hanahan, D. 1985. Heritable formation of pancreatic β-cell tumours in transgenic mice expressing recombinant insulin/simian virus 40 oncogenes. *Nature* **315:** 115.

Herbomel, P., B. Bourachot, and M. Yaniv. 1984. Two distinct enhancers with different cell specificities coexist in the regulatory region of polyoma. *Cell* **39:** 653.

Herbomel, P., S. Saragosti, D. Blangy, and M. Yaniv. 1981. Fine structure of the origin-proximal DNase I hypersensitive region in wild-type and EC mutant polyoma. *Cell* **25:** 651.

Herr, W. and Y. Gluzman. 1985. Duplications of a mutated simian virus 40 enhancer restore its activity. *Nature* **313:** 711.

Hochschild, A., N. Irwin, and M. Ptashne. 1983. Repressor structure and the mechanism of positive control. *Cell* **32:** 319.

Jongstra, J., T.L. Reudelhuber, P. Oudet, C. Benoist, C.-B. Chae, J.-M. Jeltsch, D.J. Mathis, and P. Chambon. 1984. Induction of altered chromatin structures by simian virus 40 enhancer and promoter elements. *Nature* **307:** 708.

Kenney, S., V. Natarajan, D. Strike, G. Khoury, and N.P. Salzman. 1984. JC virus enhancer-promoter active in human brain cells. *Science* **226:** 1337.

Khoury, G. and P. Gruss. 1983. Enhancer elements. *Cell* **33:** 313.

Klein, S., F. Sablitzky, and A. Radbruch. 1984. Deletion of the IgH enhancer does not reduce immunoglobulin heavy chain production of a hybridoma IgD class switch variant. *EMBO J.* **3:** 2473.

Laimins, L.A., P. Tsichlis, and G. Khoury. 1984. Multiple enhancer domains in the 3′ terminus of the Prague strain of Rous sarcoma virus. *Nucleic Acids Res.* **12:** 6427.

Levinson, B., G. Khoury, G. Vande Woude, and P. Gruss. 1982. Activation of SV40 genome by 72 base pair tandem repeats of Moloney sarcoma virus. *Nature* **295:** 568.

Luciw, P.A., J.M. Bishop, H.E. Varmus, and M.R. Capecchi. 1983. Location and function of retroviral and SV40 sequences that enhance biochemical transformation after microinjection of DNA. *Cell* **33:** 705.

Mason, J.O., G.T. Williams, and M.S. Neuberger. 1985. Transcription cell-type specificity is controlled by an immunoglobulin V_H gene promoter that includes a functional consensus sequence. *Cell* **41:** 479.

Mattaj, I.W., S. Lienhard, J. Jiricny, and E.M. De Robertis. 1985. An orientation-independent element within the U2 gene promoter acts by enabling the formation of stable RNA polymerase II transcription complexes. *Nature* (in press).

Moreau, P., R. Hen, B. Wasylyk, R. Everett, M.P. Gaub, and P. Chambon. 1981. The SV40 72 base pair repeat has a striking effect on gene expression both in SV40 and other chimeric recombinants. *Nucleic Acids Res.* **9:** 6047.

Neuberger, M.S. 1983. Expression and regulation of immunoglobulin heavy chain gene transfected into lymphoid cells. *EMBO J.* **2:** 1373.

Neuhaus, G., G. Neuhaus-Url, P. Gruss, and H.-G. Schweiger. 1984. Enhancer-controlled expression of the simian virus 40 T-antigen in the green alga *Acetabularia*. *EMBO J.* **3:** 2169.

Ott, M.O., L. Sperling, P. Herbomel, M. Yaniv, and M.C. Weiss. 1984. Tissue-specific expression is conferred by a sequence from the 5' end of the rat albumin gene. *EMBO J.* **3**: 2505.

Parslow, T.G. and D.K. Granner. 1982. Chromatin changes accompany immunoglobulin x gene activation: A potential control region within the gene. *Nature* **299**: 449.

Payne, G.S., J.M. Bishop, and H.E. Varmus. 1982. Multiple arrangements of viral DNA and an activated host oncogene in bursal lymphomas. *Nature* **295**: 209.

Picard, D. 1985. Viral and cellular transcription enhancers. In *Oxford surveys on eukaryotic genes* (ed. N. Maclean), vol. 2, p. 24. Oxford University Press, Oxford, England.

Picard, D. and W. Schaffner. 1984. A lymphocyte-specific enhancer in the mouse immunoglobulin x gene. *Nature* **307**: 80.

Queen, C. and J. Stafford. 1984. Fine mapping of an immunoglobulin gene activator. *Mol. Cell. Biol.* **4**: 1042.

Reeder, R.H. 1984. Enhancers and ribosomal gene spacers. *Cell* **38**: 349.

Rosenberg, M. and D. Court. 1979. Regulatory sequences involved in the promotion and termination of RNA transcription. *Annu. Rev. Genet.* **13**: 319.

Rosenthal, N., M. Kress, P. Gruss, and G. Khoury. 1983. BK viral enhancer element and a human cellular homolog. *Science* **222**: 749.

Ryoji, M. and A. Worcel. 1984. Chromatin assembly in *Xenopus* oocytes: In vivo studies. *Cell* **37**: 21.

Saragosti, S., G. Moyne, and M. Yaniv. 1980. Absence of nucleosomes in a fraction of SV40 chromatin between the origin of replication and the region coding for the late leader RNA. *Cell* **20**: 65.

Sassone-Corsi, P., J.P. Dougherty, B. Wasylyk, and P. Chambon. 1984. Stimulation of in vitro transcription from heterologous promoters by the simian virus 40 enhancer. *Proc. Natl. Acad. Sci.* **81**: 308.

Scheidereit, C., S. Geisse, H.M. Westphal, and M. Beato. 1983. The glucocorticoid receptor binds to defined nucleotide sequences near the promoter of mouse mammary tumour virus. *Nature* **304**: 749.

Schöler, H.R. and P. Gruss. 1984. Specific interaction between enhancer-containing molecules and cellular components. *Cell* **36**: 403.

Seguin, C., B.K. Felber, A.D. Carter, and D.H. Hamer. 1984. Competition for cellular factors that activate metallothionein gene transcription. *Nature* **312**: 781.

Sekikawa, K. and A.J. Levine. 1981. Isolation and characterization of polyoma host-range mutants that replicate in nullipotential embryonal carcinoma cells. *Proc. Natl. Acad. Sci.* **78**: 1100.

Serfling, E., M. Jasin, and W. Schaffner. 1985. Enhancers and eukaryotic gene transcription. *Trends Genet.* (in press).

Sergeant, A., D. Bohmann, H. Zentgraf, H. Weiher, and W. Keller. 1984. A transcription enhancer acts in vitro over distances of hundreds of basepairs on both circular and linear templates but not on chromatin-reconstituted DNA. *J. Mol. Biol.* **180**: 577.

Shaul, Y., W.J. Rutter, and O. Laub. 1985. A human hepatitis B viral enhancer element. *EMBO J.* **4**: 427.

Siebenlist, U., R.B. Simpson, and W. Gilbert. 1980. *E. coli* RNA polymerase interacts homologously with two different promoters. *Cell* **20**: 269.

Stewart, T.A., P.K. Pattengale, and P. Leder. 1984. Spontaneous mammary adenocarcinomas in transgenic mice that carry and express MTV/*myc* fusion genes. *Cell* **38:** 627.

Struhl, K. 1984. Genetic properties and chromatin structure of the yeast *gal* regulatory element: An enhancer-like sequence. *Proc. Natl. Acad. Sci.* **81:** 7865.

Sutcliffe, J.G., R.J. Milner, J.M. Gottesfeld, and R.A. Lerner. 1984. Identifier sequences are transcribed specifically in brain. *Nature* **308:** 237.

Swimmer, C. and T. Shenk. 1984. A viable simian virus 40 variant that carries a newly generated sequence reiteration in place of the normal duplicated enhancer element. *Proc. Natl. Acad. Sci.* **81:** 6652.

Tooze, J., ed. 1981. *Molecular biology of tumor viruses*, 2nd edition, revised: *DNA tumor viruses*. Cold Spring Harbor Laboratory, Cold Spring Harbor, New York.

Treisman, R. and T. Maniatis. 1985. The SV40 enhancer increases the number of RNA polymerase II molecules on linked DNA. *Nature* **315:** 72.

Treisman, R., M.R. Green, and T. Maniatis. 1983. *cis* and *trans* activation of globin gene transcription in transient assays. *Proc. Natl. Acad. Sci.* **80:** 7428.

Vasseur, M., M. Katinka, P. Herbomel, M. Yaniv, and D. Blangy. 1982. Physical and biological features of polyoma virus mutants able to infect embryonal carcinoma cell lines. *J. Virol.* **43:** 800.

Wabl, M.R. and P.D. Burrows. 1984. Expression of immunoglobulin heavy chain at a high level in the absence of a proposed immunoglobulin enhancer element in *cis*. *Proc. Natl. Acad. Sci.* **81:** 2452.

Walker, M.D., T. Edlund, A.M. Boulet, and W.J. Rutter. 1983. Cell-specific expression controlled by the 5'-flanking region of insulin and chymotrypsin genes. *Nature* **306:** 557.

Wasylyk, B., C. Wasylyk, P. Augereau, and P. Chambon. 1983. The SV40 72 bp repeat preferentially potentiates transcription starting from proximal natural or substitute promoter elements. *Cell* **32:** 503.

Weber, F. and W. Schaffner. 1985a. Simian virus 40 enhancer increases RNA polymerase density within the linked gene. *Nature* **315:** 75.

———. 1985b. Enhancer activity correlates with the oncogenic potential of avian retroviruses. *EMBO J.* **4:** 949.

Weber, F., J. de Villiers, and W. Schaffner. 1984. An SV40 "enhancer trap" incorporates exogenous enhancers or generates enhancers from its own sequences. *Cell* **36:** 983.

Weiher, H., M. König, and P. Gruss. 1983. Multiple point mutations affecting the simian virus 40 enhancer. *Science* **219:** 626.

Wildeman, A.G., P. Sassone-Corsi, T. Grundström, F. Zenke, and P. Chambon. 1984. Stimulation of in vitro transcription from the SV40 early promoter by the enhancer involves a specific *trans*-acting factor. *EMBO J.* **3:** 3129.

Yaniv, M. 1984. Regulation of eukaryotic gene expression by *trans*-activating proteins and *cis*-acting DNA elements. *Biol. Cell.* **50:** 203.

Interactions between the SV40 Early Promoter and Cellular Proteins

A.G. Wildeman, M. Zenke, C. Schatz, K. Takahashi, H. Barrera-Saldana, T. Grundstrom, M. Wintzerith, H. Matthes, M. Vigneron, and P. Chambon

Laboratoire de Génétique Moléculaire des Eucaryotes du CNRS
Unité 184 de Biologie Moléculaire et de Génie Génétique de l'INSERM
Institut de Chimie Biologique, Faculté de Médecine
67085 Strasbourg, France

The DNA sequences required for early transcription of SV40 have been shown to include, in addition to a TATA box, both an upstream sequence element, the 21-bp repeat region (e.g., Baty et al. 1983), and an enhancer element, the 72-bp repeat region (Banerji et al. 1981; Benoist and Chambon 1981; Gruss et al. 1981; Moreau et al. 1981). Efficient transcription requires the presence of both of these elements, although the enhancer is unique in that it can function in either orientation over distances of several kilobase pairs (e.g., Wasylyk et al. 1983). Using in vitro transcription systems, it has been shown that there are unique, specific factors that are responsible for the activities of the 21-bp regions (the factor Sp1; Dynan and Tjian 1983) and the enhancer (Wildeman et al. 1984; Sassone-Corsi et al. 1985).

To define precisely the nucleotides within these two elements that are required for their function, we have used site-directed, mismatched-primer mutagenesis to introduce point-mutation clusters throughout both and analyzed the mutant phenotypes by transient expression of the recombinant plasmids into HeLa cells, using quantitative S1 nuclease analysis of the transcribed RNA. We have then made use of the observation that DNase I footprinting, carried out in a nuclear extract that is transcriptionally active, can be used to analyze the binding of the 21-bp- and enhancer-specific factors and to assess directly the effect of point mutations on factor binding in vitro.

The SV40 21-bp repeat contains six GC-rich motifs, each six nucleotides in length. There are two such motifs in each of the two 21-

bp perfect repeats as well as in the 22-bp imperfect repeat. They are termed GC-I to GC-VI, with the GC-I being adjacent to the TATA box region of the early promoter and GC-VI being adjacent to the 72-bp enhancer region. We have assessed the role of each GC motif in viral early transcription by mutating them individually, substituting A-T base pairs for three of the six G-C base pairs within each block. We find that all GC motifs contribute to the efficiency of the early promoter, but to differing degrees. Mutations in GC-I and GC-VI lower transcription to 4.8%, 20.0%, 15.0%, 42.0%, 33.0%, and 56.0% of wild type, respectively, and within each of the three 21- or 22-bp sequences the GC motif nearest the TATA box appears to play the more important role. Recombinants having both GC motifs in each repeat simultaneously mutated were also constructed, and analysis of their phenotypes showed that for early transcription the 22-bp sequence, which is nearest the TATA box, is the most important component of the 21-bp repeat region. The 21-bp sequence nearest the enhancer has the least influence on the early promoter. In all cases, the double-mutated templates exhibited lower transcription activity than either of the recombinants with a mutation in only one of the two GC motifs in that repeat.

The in vivo phenotype of these GC mutants is essentially reproduced in vitro by using a HeLa nuclear extract as previously described (Wildeman et al. 1984). Furthermore, as shown in Figure 1, using increasing amounts of this total nuclear extract, specific sequences of the SV40 promoter region are protected from DNase I attack. On both the early and late coding strands, the entire 21-bp repeat region is strongly protected (E7 and L8), although specific nucleotides in GC-I and GC-IV on the late strand and GC-I on the early strand are somewhat accessible to digestion. The mutation of any GC motif of the 22-bp sequence resulted in deprotection of it only, and templates having both GC motifs in any of the three 21-bp or 22-bp repeats mutated retained protection, although slightly diminished, over the remaining four GC motifs (not shown). Thus, although there may be some cooperativity in factor binding among the various GC motifs, each one appears to be capable of factor binding. If each is binding one molecule of Sp1, albeit with different efficiencies, the observation that multiple GC motif mutations are more deleterious for transcription than single mutations suggests that it is the multiple DNA-protein complexes that maximize the strength of the SV40 early promoter.

The nucleotide sequences required for the function of the SV40 enhancer were similarly assessed by generating a series of point mutations scanning an enhancer that contained a single 72-bp se-

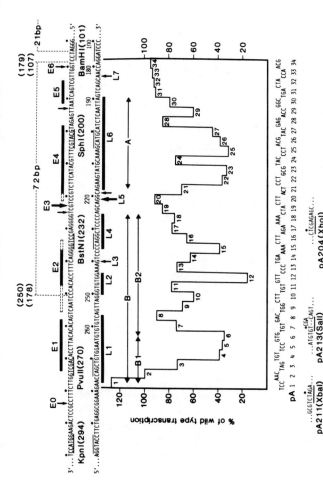

Figure 1 Comparison of the locations of regions of DNase I protection and hypersensitivity observed in vitro with the results of the in vivo activity of enhancer point mutations. Regions of protection are indicated as solid lines on the sequence, and sites of hypersensitivity as arrows. Below the sequence is shown the profile of the effect in vivo on transcription (as determined by calcium phosphate transfection of HeLa cells and quantitative S1 nuclease analysis of RNA transcribed from the early promoter) of point mutations at different positions throughout the enhancer. The nucleotide changes present in each of the 34 mutated templates [pA1–pA34] are indicated below the profile, as are the positions of XbaI, SalI, and XhoI sites, which were introduced to facilitate the construction of insertion and inversion mutants.

21

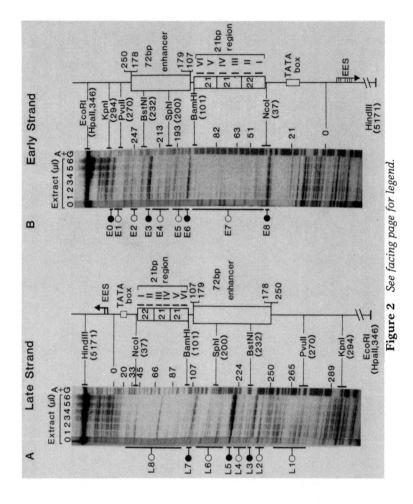

Figure 2 *See facing page for legend.*

22

Figure 2 (*see facing page*) DNase footprinting of the SV40 promoter regions in nuclear extracts. Promoter DNA fragments were labeled on the early or late coding strands at the *Hin*dIII or *Eco*RI sites, respectively (for description of recombinant plasmid used, see Wildeman et al. 1984), and purified by gel electrophoresis. Footprinting reactions were carried out using 0–6 µl of HeLa cell nuclear extract and ionic conditions required for enhancer-dependent transcription (see Wildeman et al. 1984). Extracts were preincubated with pBR322 DNA, then with labeled template for 10 min at 20°C. DNase I was added at an empirically determined concentration for 90 sec, and reactions were terminated by SDS-phenol extraction. Samples were analyzed on salt-gradient 8% polyacrylamide gels. To the right of each autoradiograph is shown the positions of the enhancer and 21-bp repeat regions, and to the left is indicated the regions whose sensitivity to DNase I is increased (●) or decreased (○) after incubation in the nuclear extract.

23

quence. Each of the 34 mutations altered clusters of three nucleotides, and the actual nucleotide changes and in vivo phenotype of the mutated templates are shown in Figure 1. This analysis revealed three regions of the enhancer that tolerated poorly the introduction of mutations, with one lying upstream of the 72-bp sequence. The regions include approximately nucleotides 185–220, 225–257, and 260–276. A recombinant (pA94, not shown) having nucleotides 202–204, 245–249, and 268–272 simltaneously mutated is almost totally inactive for transcription from the early promoter ($<1\%$ of wild type), reinforcing the importance of these three regions. In vitro DNase footprinting over the enhancer in the same nuclear extracts as above (which exhibit some in vitro enhancer-dependent transcription; Wildeman et al. 1984; Sassone-Corsi et al. 1985) identifies particular sequences that interact with proteins in the extract (Fig. 2). These are indicated as L1, L2, L4, and L6 on the late strand and E1, E2, and E5 on the early strand. Simultaneous with this protection is the appearance of DNase I hypersensitivity at L3, L5, E0, and E3. Both this protection and hypersensitivity appear to result from the specific *trans*-acting factor(s) previously shown to be responsible for the enhancer activity in vitro, since they are lost when a competitor fragment extending from *Bam*HI to *Kpn*I is added to the footprinting reaction (not shown). Neither a mutated competitor fragment prepared from the pA94 template (see above) nor pBR322 DNA competes for the footprint.

The DNase I footprinting results have been superimposed onto the SV40 enhancer sequence drawn above the scanning mutant profile in Figure 1. The regions of protection lie over those portions of the enhancer that tolerate poorly the presence of point mutations, whereas the hypersensitive sites are located at positions where mutations have little or no effect on transcription. This correlation lends support to our previous suggestion that the in vitro effect of the enhancer on transcription is a true reflection of at least part of the enhancer activity in vivo. Although it is not known how these specific DNA-protein interactions function to direct transcription to either homologous or heterologous promoters, the in vitro footprinting assay enabled a further dissection of the protein-binding domains in the SV40 enhancer. Using point-mutated templates in the footprinting reaction, it is found that the regions defined by L1 and E1 (domain B1), by L2, L4, and E2 (domain B2), and by E4 and L6 (domain A) appear to each be separate domains. Point mutations in domain A which are "down" in transcription lead to a deprotection of residues throughout that domain only and have no effect on binding in B1 and B2 (not shown). Furthermore, as many as 90 nucleotides can be inserted at an artificial *Xho*I site created between domains A and B (see Fig. 2) without affecting protein binding or abolishing transcriptional activity (not shown). Whereas domain B appears to consist of two separate domains, point mutations in B1 do affect slightly the binding over B2, and the effect of inserting 4 nucleotides at an artificial *Sal*I site near the boundary of B1 and B2 is to significantly lower transcriptional activity in vivo and to deprotect nucleotides up to and including 263. The point mutations in B1 also cause deprotection of residues that include 260–263, and so we cannot yet conclude that B1 and B2 are distinct from one another and perhaps recognized by different proteins. In any case there are almost certainly proteins unique for domains A and B; this is further supported by the observations that either A or B domains can be inverted without suppressing enhancer activity and that the inverted sequences continue to be specifically protected against DNase I digestion regardless of the orientation of the adjacent domain (not shown).

Taken together, the in vivo and in vitro results suggest that the SV40 enhancer consists of multiple elements, each recognized by specific components of the transcription machinery and that it is the simultaneous functioning of all of these elements that maximizes enhancer activity. Further purification of these components will be necessary for assessing their mechanism of action.

REFERENCES

Banerji, J., S. Rusconi, and W. Schaffner. 1981. Expression of a β-globin gene is enhanced by remote SV40 DNA sequences. *Cell* **27:** 299.

Baty, D., H. Barrera-Saldana, R. Everett, M. Vigneron, and P. Chambon. 1983. Mutational dissection of the 21 bp repeat region of the SV40 early promoter reveals that it contains overlapping elements of the early-early and late-early promoters. *Nucleic Acids Res.* **12:** 915.

Benoist, C. and P. Chambon. 1981. The SV40 early promoter region: Sequence requirements *in vivo*. *Nature* **290:** 304.

Dynan, W.S. and R. Tjian. 1983. The promoter-specific transcription factor Sp1 binds to upstream sequences in the SV40 early promoter. *Cell* **35:** 79.

Gruss, P., R. Dhar, and G. Khoury. 1981. Simian virus 40 tandem repeated sequences as an element of the early promoter. *Proc. Natl. Acad. Sci.* **78:** 943.

Moreau, P., R. Hen, B. Wasylyk, R. Everett, M.P. Gaub, and P. Chambon. 1981. The SV40 72 base pair repeat has a striking effect on gene expression both in SV40 and other chimeric recombinants. *Nucleic Acids Res.* **9:** 6047.

Sassone-Corsi, P., A. Wildeman, and P. Chambon. 1985. A *trans*-acting factor is responsible for the simian virus 40 enhancer activity *in vitro*. *Nature* **313:** 458.

Wasylyk, B., C. Wasylyk, P. Augereau, and P. Chambon. 1983. The SV40 72 bp repeat preferentially potentiates transcription starting from proximal natural or substitute promoter elements. *Cell* **32:** 503.

Wildeman, A.G., P. Sassone-Corsi, T. Grundstrom, M. Zenke, and P. Chambon. 1984. Stimulation of *in vitro* transcription from the SV40 early promoter by the enhancer involves a specific *trans*-acting factor. *EMBO J.* **3:** 3129.

Sequence Duplications That Reactivate Mutated SV40 Enhancer Elements

W. Herr, J. Clarke, and Y. Gluzman

Cold Spring Harbor Laboratory
Cold Spring Harbor, New York 11724

The mechanism by which enhancers exert their effects is not known; even the nucleotide sequences required have until recently been poorly defined. There are no strong homologies between the sequences of different enhancers, but a number of short, degenerate consensus sequences have been identified, including (1) the "core" element GTGG,A/T,A/T,A/T,G (Laimins et al. 1982; Weiher et al. 1983) and (2) stretches of alternating purines and pyrimidines (Pu/Py) (Nordheim and Rich 1983). In the prototype SV40 enhancer there is a tandem duplication of 72 bp, with each 72-bp sequence containing both a core element and an 8-bp Pu/Py stretch. Just upstream of the 72-bp duplication lies a second 8-bp Pu/Py sequence. We have recently described an SV40 enhancer mutant that contains a single 72-bp element and base substitutions in both of the Pu/Py segments (Herr and Gluzman 1985). Analysis of revertants derived from this enhancer mutant suggested that duplications of the core element could compensate for the loss of function caused by the mutations in the Pu/Py elements. Here we review those results and, in addition, describe the structure of revertants of an SV40 enhancer mutant containing an altered core element.

RESULTS AND DISCUSSION

The top of Figure 1 shows the structure of the Pu/Py mutant *dpm*12, which contains two transversion point mutations within each of the two Pu/Py elements (hatched boxes a and b). These combined mutations nearly abolish the ability of SV40 to grow in CV-1 cells and impair the ability of the enhancer to stimulate transcription of the human β-globin gene, as measured in a transient expression assay in HeLa cells. By transfection of large amounts of *dpm*12 SV40 DNA into CV-1 cells, followed by serial passage of the resulting virus stocks, we were able to isolate growth revertants. Figure 1 shows the structure of the enhancer region from each of 18 independent

27

revertants; each revertant contains a tandem duplication of between 45 bp and 135 bp of the mutated enhancer region. (The extents of the duplicated regions are indicated in Fig. 1 by open rectangular boxes.) The revertant phenotypes cosegregate with these duplications, and the duplications restore enhancer function in the HeLa cell transient expression assay (Herr and Gluzman 1985).

Examination of the new sequences created at the junctions of the different tandem duplications indicates that these new sequences

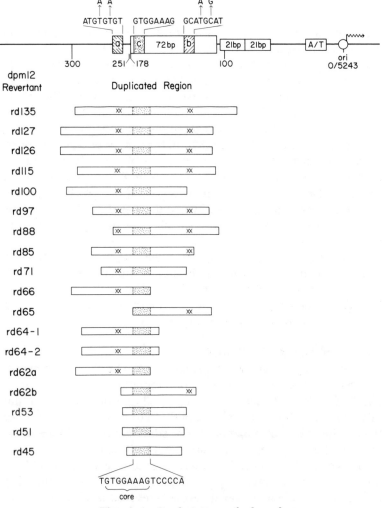

Figure 1 *See facing page for legend.*

per se are not responsible for the revertant phenotypes, because no homologies between these sequences are apparent. Instead, the consistent duplication of a 15-bp region (the stippled areas in Fig. 1) in all of the revertants suggests that this latter region, which spans the core element (see the bottom of Fig. 1), plays a critical role in restoring activity to the *dpm*12 mutant. These results indicate that the SV40 enhancer is composed of at least three functional *cis*-acting elements – the two Pu/Py regions (both sets of Pu/Py mutations impair enhancer function independently [W. Herr and Y. Gluzman, unpubl.]) and the core region – and that duplication of one element, the core region, can compensate for loss of function in the other two elements. Moreover, the core sequences appear to act independently of the Pu/Py elements, because enhancer function is restored when the core is duplicated but the Pu/Py regions remain mutated. These results left unanswered, however, the question of whether the three regions are functionally equivalent; i.e., Can the regions containing the Pu/Py elements, in turn, compensate for loss of function in the core element? To address this question, we have constructed an SV40 mutant in which the core element is mutated, and we have selected for revertants of this mutant by selection for improved growth in CV-1 cells (as described for the selection of *dpm*12 revertants).

The core mutant *dpm*6 contains a single 72-bp element with two transversions that alter the core sequence from GTGGAAAG to GTCCAAAG, as shown at the top of Figure 2. As with *dpm*12, SV40

Figure 1 (*see facing page*) Regions tandemly duplicated in revertants of the SV40 enhancer mutant *dpm*12. (*Top*) The elements within an SV40 control region containing only one copy of the 72-bp element. Right to left: the early transcriptional start site (wavy arrow); the origin of replication (ori); the AT-rich TATA-like element (A/T); the GC-rich 21-bp repeats; and the single 72-bp element. Numbers below the diagram refer to nucleotide positions in the SV40 sequence. Locations of the Pu/Py segments (boxes a and b) and the core element (box c) are shown with the respective wild-type sequences displayed above each box; the four *dpm*12 transversions are indicated above the two Pu/Py segments. (*Bottom*) The extents of the regions tandemly duplicated within each of the 18 *dpm*12 revertants are indicated by open rectangular boxes. Revertants are aligned from top to bottom in order of size, with the sizes indicated by revertant designation (e.g., *rd*135 is a 135-bp long revertant duplication). The precise endpoints of each duplication are described elsewhere (Herr and Gluzman 1985). (× ×) Positions of the original point mutations in the duplication. The stippled area within each rectangular box corresponds to the 15-bp region common to all of the duplications; the sequence of this region is shown at bottom with a brace indicating the core element (Weiher et al. 1983).

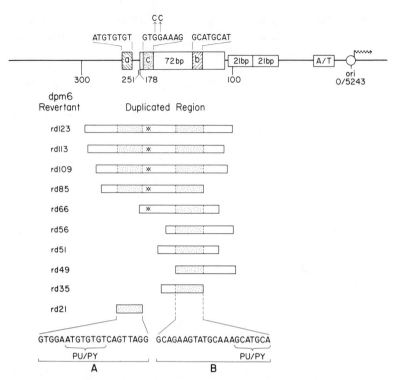

Figure 2 Regions duplicated in revertants of the SV40 enhancer mutant *dpm*6 with a mutated core element. (*Top*) Diagram of the SV40 early promoter (see Fig. 1). The two *dpm*6 transversions are shown above the sequence of the core element (box c). (*Bottom*) The extents of the duplicated regions in the *dpm*6 revertants are indicated by rectangular boxes. (× ×) Sites of the point mutations when duplicated. The stippled regions within the duplicated sequences represent the two regions that are duplicated, singly or together, in the various *dpm*6 revertants. These are referred to as the A and B boxes, and their sequences are shown at bottom with the Pu/Py sequences indicated by braces.

virus containing the *dpm*6 mutations grows poorly in CV-1 cells, and the *dpm*6 enhancer shows a reduced capacity to stimulate transcription of the human β-globin gene. Figure 2 shows the structure of the enhancer region from 10 independent *dpm*6 revertants; like the *dpm*12 revertants, these enhancers also have tandem duplications of the mutated enhancer. We have not yet shown that the *dpm*6 duplications are directly responsible for the revertant phenotypes, but this result is likely because of the close analogy to the *dpm*12 duplications, all of which restored enhancer activity.

The structures of the *dpm*6 revertants indicate that duplication of at least two separate regions can compensate for the mutated core element; these two regions are shown in Figure 2 as stippled boxes within the duplicated sequences. These regions are referred to as the A and B boxes, and the sequences are displayed at the bottom of Figure 2. The A box spans the Pu/Py sequence outside of the 72-bp element, and the B box contains 7 out of 8 bp of the other Pu/Py sequence. Four of the *dpm*6 duplications (*rd*85, *rd*109, *rd*113, *rd*123) span both boxes; revertants *rd*35 through *rd*66 span the B box but not the A region; and one revertant, *rd*21, contains only the A box. These results suggest that either of the regions containing the sequences mutated in *dpm*12 is able to compensate independently for the mutated core element.

We have also constructed SV40 mutants that contain double transversions in both the core element and the Pu/Py sequences within either the A or B box; these new double mutants are called *dpm*16 and *dpm*26, respectively. Four revertants of *dpm*16 have so far been examined; all four duplicate the B box but not necessarily the mutated A or core ("C" box) regions. Two revertants of *dpm*26 have been sequenced, and they have structures complementary to the *dpm*16 revertants: The A region is duplicated, but the mutated B and C boxes are not always duplicated. Taken together with the *dpm*12 and *dpm*6 revertant structures, these results suggest that the SV40 enhancer indeed contains three functionally equivalent regions–boxes A, B, and C–each of which is able to compensate by duplication for loss of function within the other regions.

Both in vivo and in vitro enhancer competition experiments suggest that *trans*-acting factors are involved in enhancer function (Schöler and Gruss 1984; Sassone-Corsi et al. 1985). The three elements described here, boxes A–C, may represent binding sites for one or more *trans*-acting factors. If so, then one model to explain reactivation of enhancers by duplication of separate elements would postulate that the same *trans*-acting factor interacts with each of the three elements, and that for the enhancer to function efficiently, a certain number of sites must be occupied by the factor. Consistent with this model, both the A and C boxes share homology to the core consensus sequence, which may represent such a factor-binding site. The B box, however, does not share this homology (the best match to the core consensus sequence is 6 out of 8 nucleotides in the sequence ATGCAAAG). The B box may represent instead the binding site of a factor with a different sequence specificity from that of the putative core binding factor. Nevertheless, the B box can compensate for mutations in the A and C boxes, suggesting that any

B-box-specific binding factor stimulates transcription in a manner analogous to the putative factors interacting with the A and/or C boxes. This interpretation argues that *trans*-acting enhancer factors have at least two distinct functional domains: a domain that stimulates transcription that is common to the different factors and a DNA-binding domain that differs between separate factors. Such separate domains for DNA binding and activation of transcription have been previously described in the repressors for *E. coli* lambdoid phages (Wharton et al. 1984; Hochschild et al. 1983). These results, then, raise the possibility that the SV40 enhancer is regulated by two (or more?) factors that coordinately function to stimulate transcription.

ACKNOWLEDGMENTS
We thank our colleagues at Cold Spring Harbor Laboratory for many helpful discussions. This research was supported by U.S. Public Health Service grant CA-13106 from the National Cancer Institute.

REFERENCES
Herr, W. and Y. Gluzman. 1985. Duplications of a mutated simian virus 40 enhancer restore its activity. *Nature* **313:** 711.
Hochschild, A., N. Irwin, and M. Ptashne. 1983. Repressor structure and the mechanism of positive control. *Cell* **32:** 319.
Laimins, L.A., G. Khoury, C. Gorman, B. Howard, and P. Gruss. 1982. Host-specific activation of transcription by tandem repeats from simian virus 40 and Moloney murine sarcoma virus. *Proc. Natl. Acad. Sci.* **79:** 6453.
Nordheim, A. and A. Rich. 1983. Negatively supercoiled simian virus 40 DNA contains Z-DNA segments within transcriptional enhancer sequences. *Nature* **303:** 674.
Sassone-Corsi, P., A. Wildeman, and P. Chambon. 1985. A *trans*-acting factor is responsible for the simian virus 40 enhancer activity *in vitro*. *Nature* **313:** 458.
Schöler, H.R. and P. Gruss. 1984. Specific interaction between enhancer-containing molecules and cellular components. *Cell* **36:** 403.
Weiher, H., M. Konig, and P. Gruss. 1983. Multiple point mutations affecting the simian virus 40 enhancer. *Science* **219:** 626.
Wharton, R.P., E.L. Brown, and M. Ptashne. 1984. Substituting an α-helix switches the sequence-specific DNA interactions of a repressor. *Cell* **38:** 361.

The Polyoma Virus Enhancer: Multiple Sequence Elements Required for Transcription and DNA Replication

J.A. Hassell, C.R. Mueller, and W.J. Muller
Department of Microbiology and Immunology, McGill University
Montreal, Quebec, Canada H3A 2B4

The DNA between the transcriptional start sites for the early and late mRNAs of polyoma virus (Py) contains numerous sequence elements important for transcription and DNA replication. These include the early and late promoters and the functional origin for DNA replication (ori). We are interested in defining the sequences of these *cis*-acting regulatory elements and in identifying and purifying the viral and cellular proteins that interact with them to elucidate the mechanism of initiation of transcription and DNA replication. To this end, we have begun to map by in vitro mutagenesis techniques the locations of these various regulatory elements within the Py noncoding region and to study the interaction of viral and cellular proteins with these sequences (Muller et al. 1983; Pomerantz et al. 1983; Mueller et al. 1984; Pomerantz and Hassell 1984).

The early promoter of Py, like that of simian virus 40 (SV40), is made up of multiple sequence elements. These include a TATA box–cap site element, a functionally redundant middle element analogous to the SV40 21-bp repeats, and an enhancer (Jat et al. 1982; Mueller et al. 1984). The borders of the Py enhancer are not known in detail, but it is known that the enhancer is contained within a 226-bp fragment that maps between nucleotides (nt) 5039 and 5265 or positions −400 and −175 relative to the major start sites for the early mRNAs (Tyndall et al. 1981; de Villiers and Schaffner 1981; Jat et al. 1982; Mueller et al. 1984). Two functionally redundant elements that form part of the Py ori are also situated within the enhancer borders. The function of these elements, which have been named α and β, is not known, but one or the other element must be coupled to another ori element, termed the core, to constitute a functional origin for DNA replication (Muller et al.

1983). The core maps outside of the enhancer region and it is likely the site of initiation of viral DNA replication.

To determine whether the sequences required for enhancer function and those required for replication were the same or not, we defined their borders by deletion mutagenesis. The enhancer is composed of three sequence elements that we have named elements 1, 2, and 3. Pairs of these elements act an enhancers, but each individual element is relatively inert as an enhancer. Enhancer elements 2 and 3 contain the α and β replication elements within their borders, and it is very likely that the same sequences in each element play a dual role in transcription and DNA replication.

Location of the Polyoma Virus Enhancer

To facilitate mapping the Py enhancer, we substituted the Py enhancer-promoter region for that of SV40 in the plasmid pSV2cat to generate a new plasmid, named pdPyEcat. In brief, this plasmid is made up of the Py early enhancer-promoter region (Py nt 4632 to 154, or −810 to +4) joined to the chloramphenicol acetyltransferase (CAT) gene, which is followed by an SV40–β-globin DNA segment that provides splicing, transcription termination, and polyadenylation signals (Gorman et al. 1982). pdPyEcat and its deleted derivates were transfected into NIH-3T3 cells, and the amount of CAT activity obtained from the transfected cells was used as an indirect measure of expression of the CAT gene (cat).

To map the Py enhancer, we isolated a set of 5′ and 3′ unidirectional deletion mutants with lesions within the enhancer, measured their capacity to express cat after transfection of mouse 3T3 cells, and compared this activity with that obtained after transfection of these same cells with pdPyEcat. Analysis of 10 5′ unidirectional deletion mutants revealed that the 5′ border of the enhancer was between nt positions 5075 (88% of wild type) and 5113 (9% of wild type). Two mutants with deletions between these endpoints expressed cat at intermediate levels (∼30% of wild type).

To map the 3′ border of the enhancer, we began mutagenesis at nt 90 (−60 in pdPyEcat) and extended the deletion in the upstream direction. All of these deletion mutants retain the TATA box–cap site element downstream of nt 90. The results of characterizing these mutants revealed that sequences up to nt 5130 could be removed without affecting expression of cat. However, deletion of sequences upstream of position 5130 to 4947 abolished the capacity of the plasmid to express cat. Taken together, these results suggested that the Py enhancer was contained within a 55-bp fragment located between nt positions 5075 and 5130 (−367 to −310).

34

To determine whether this 55-bp fragment possessed enhancer activity, we cloned it alone in its natural orientation before the TATA box–cap site element of Py in the *cat* vector. Transfection of this plasmid into 3T3 cells yielded unexpectedly low levels of CAT activity (at best 15% of the wild-type plasmid). This result indicated that the enhancer was not contained within the 55-bp fragment. By analyzing a variety of additional deletion mutants, we were able to show that two overlapping sets of sequences, namely sequences between nt 5039 and 5130 and those between nt 5075 and 5229, could function independently to effect full expression of *cat* coupled to the Py promoter (TATA box–cap site element). These sequences have in common the 55-bp DNA fragment located between nt positions 5075 and 5130.

These results are compatible with two alternative structures for the Py enhancer. One possibility is that the enhancer contains a central core (nt 5075–5130) and two flanking, functionally redundant elements located between nt 5039 and 5075 or between nt 5130 and 5229. Alternatively, the enhancer could be composed of three independent sequence modules that function in pair-wise combinations. To distinguish between these models, we deleted the central element (between nt 5092 and 5130) from a modified Py-*cat* plasmid. The latter contained Py sequences from nt position 5039 to 5092, the enhancer-internal deletion described above, and Py DNA from nt position 5130 to 154. This plasmid expressed *cat* at essentially wild-type levels (60% of wild type). These results are consistent with the second model proposed above and not with the first.

To confirm that the Py enhancer is composed of multiple sequence elements that function in pair-wise combinations, we cloned various Py DNA segments in another *cat* plasmid, named pTE1. This plasmid contains a hybrid transcription unit composed of the thymidine kinase (*tk*) promoter (−109 to +50) fused to *cat* and accompanying downstream signals from SV40 and rabbit β-globin to effect proper processing of its transcript. Py enhancer fragments were cloned at a site some 600 bp upstream from the start site for *cat* transcription in both possible orientations. A total of 14 different plasmids were assayed for their capacity to express *cat* after transfection of mouse cells by comparison with the parental plasmid, pTE1. The results of several experiments are summarized in Table 1. As a general rule, single enhancer elements could not activate *cat* expression. One notable exception was element 2, which could activate *cat* expression in one orientation (−), but not in the other (+). By contrast, pairs of enhancer elements were nearly as effective as

Table 1
Elevation of *cat* Expression by Polyoma Virus Enhancer Elements

Py DNA segment	Enhancer element	Relative orientation	Enhancement factor
5039–5092	1	+	1
5039–5092	1	–	1
5075–5130	2	+	1
5075–5130	2	–	10
5130–5265	3	+	3
5130–5265	3	–	3
5039–5130	1 + 2	+	13
5039–5130	1 + 2	–	78
5039–5092/5130–5265	1 + 3	+	40
5039–5092/5130–5265	1 + 3	–	40
5075–5229	2 + 3	+	60
5075–5229	2 + 3	–	77
5039–5265	1 + 2 + 3	+	80
5039–5265	1 + 2 + 3	–	100

all three elements together in augmenting *cat* expression. To insure that the enhanced levels of *cat* expression mediated by the Py enhancer elements were the result of increased transcription of *cat*, we measured the abundance and 5′ terminus of the *cat* mRNA by S1 nuclease analysis after transfection of 3T3 cells with various plasmids. These data showed that increased levels of correctly initiated transcripts resulted after inclusion of pairs of Py enhancer elements in pTE1.

Polyoma Virus Replication Activator Elements

We mapped the α and β replication elements, using methods described previously (Muller et al. 1983). The α element maps within the borders of enhancer element 2, between nt 5097 and 5126, whereas the β element maps within the borders of enhancer element 3, between nt 5172 and 5202. Furthermore, deletions that inactivate the α function also abolish the activity of enhancer element 2, and deletions that inactivate the β function simultaneously cripple enhancer element 3. Although more mutants will have to be isolated to prove unequivocally that the same sequences in each element mediate transcription enhancement and replication activation, all the data gathered so far support this contention.

Because the replication activators and the enhancer elements appeared to be coincident, we were curious to learn whether the α and β elements, like enhancers, could activate replication independent of their orientation or position relative to the core element.

We measured the replication of a number of recombinant plasmids in which the orientation and position of the α and β elements were altered relative to the core and found that α and β could separately activate replication independent of their orientation. However, these elements could not activate the core for replication when they were moved either separately or together to a site approximately 200 bp from the late border of the core, nor could they function when placed 50 bp from the early border of the core. It is noteworthy that the β element is normally situated at least 60 bp from the late border of the core, whereas the α element is located about 135 bp from this same border in the viral genome. These characteristics of the replication elements contrast with those of enhancers.

DISCUSSION

The locations of the enhancer and replication elements within the Py genome are shown in Figure 1 together with other features resident in this area. Each enhancer element contains an inverted repeat and a sequence homologous to the SV40 or adenovirus E1A enhancer core sequence. Elements 1 and 3 contain an SV40 enhancer core sequence, whereas element 2 contains an adenovirus E1A enhancer core sequence. The α and β elements are located within the borders of elements 2 and 3, respectively, and the enhancer core sequences as well as the inverted repeats lie within α and β. These are two DNase I hypersensitive sites within the enhancer region (Herbomel et al. 1981). These sites map within enhancer elements 2 and 3, immediately adjacent to the α and β replication elements.

The Py enhancer consists of at least three elements. Individual elements are incapable of acting as enhancers, but pairs of elements function to augment gene expression nearly as well as all three elements together do. By contrast, either α or β will independently activate the Py core for DNA replication. This occurs independent of their orientation relative to the core, but replication activation only occurs when the α and β elements are positioned near the late border of the core (within 50–60 bp). Moreover, replication cannot be activated if both elements are moved together from the late core border. This is so even though such a fragment bearing α and β is capable of acting as an enhancer of transcription.

The results we presented here are consistent with the existence of viable viruses with deleted genomes. For example, the virus dl2039L lacks sequences between nt 5100 and 5131 yet is viable (Cowie et al. 1981). This virus does not contain enhancer element 2 and the α replication element (Fig. 1). Similarly, another virus (dl1024) lacks enhancer element 3 and the β element (Luthman et

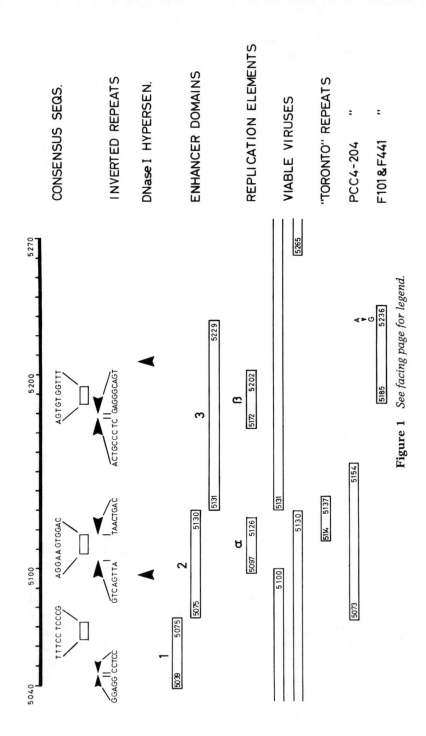

Figure 1 *See facing page for legend.*

Figure 1 Schematic of features within the Py enhancer region. The solid, numbered line represents noncoding Py DNA. The consensus sequences shown below the line represent those sequences in Py DNA that are homologous to the enhancer core sequence found in SV40 (–G/C,G/C,C/T,GTGG,A/T,A/T,A,T) and Ad5 (A/C,GGAAGTG,A/C). The inverted repeats referred to in the text are also shown adjacent to the consensus sequences. The two arrowheads pointing upward depict the sites of DNase I hypersensitivity found in Py DNA in vivo. The outer limits of the enhancer domains are shown, as are those of the replication activator elements α and β. The sequence retained by two viable viruses of Py in this area of the genome are shown. One virus lacks sequences between nt 5100 and 5131, whereas the other is deleted of sequences between nt 5130 and 5265. Those sequences that are commonly duplicated in various strains (i.e., the Toronto strain) and host-range mutants of Py are represented by the three boxed segments at the bottom. One mutant, F441, contains only a base substitution (A to G) at nt 5230 and no duplication, whereas the mutant F101 contains the same substitution as well as a duplication of those sequences between nt 5185 and 5236.

39

al. 1982). Moreover, there are numerous strains and host-range mutants of Py whose genomes contain duplications of enhancer elements 2 and 3, or α and β, as well as mutations within these elements. Duplications may increase the strength of the elements, whereas mutations may alter their specificity, thereby allowing for their interaction with proteins found only in embryonic cells or in particular differentiated cells in the mouse. In this regard it is noteworthy that the α element fails to activate the Py core for replication in early mouse embryos, whereas the β element does (D.O. Wirak et al., in prep.).

Finally, it is interesting that the enhancer elements and replication elements, which are likely the same sequences, act differently to effect transcription and DNA replication. Whether this reflects a mechanistic difference or not remains to be determined.

REFERENCES

Cowie, C., C. Tyndall, and R. Kamen. 1981. Sequences at the capped 5'-ends of polyoma virus late region mRNAs: An example of extreme terminal heterogeneity. *Nucleic Acids Res.* **9:** 6305.

de Villiers, J. and W. Schaffner. 1981. A small segment of polyoma virus DNA enhances the expression of a cloned β-globin gene over a distance of 1,400 base pairs. *Nucleic Acids Res.* **9:** 6251.

Gorman, C.M., L.F. Moffat, and B.H. Howard. 1982. Recombinant genomes which express chloramphenicol acetyltransferase in mammalian cells. *Mol. Cell. Biol.* **2:** 1044.

Herbomel, P., S. Saragosti, D. Blangy, and M. Yaniv. 1981. Fine structure of origin-proximal DNase I–hypersensitive region in wild-type and E.C. mutant polyoma. *Cell* **25:** 651.

Jat, P., U. Novak, A. Cowie, C. Tyndall, and R. Kamen. 1982. DNA sequences required for specific and efficient initiation of transcription at the polyoma virus early promoter. *Mol. Cell. Biol.* **2:** 737.

Luthman, H., M.G. Nilsson, and S. Magnusson. 1982. Non-contiguous elements of the polyoma gene required in *cis* for DNA replication. *J. Mol. Biol.* **161:** 533.

Mueller, C.R., A.-M. Mes-Masson, M. Bouvier, and J.A. Hassell. 1984. Location of sequences in polyomavirus DNA that are required for early gene expression *in vivo* and *in vitro*. *Mol. Cell. Biol.* **4:** 2594.

Muller, W.J., C.R. Mueller, A.-M. Mes, and J.A. Hassell. 1983. Polyomavirus origin for DNA replication comprises multiple genetic elements. *J. Virol.* **47:** 586.

Pomerantz, B.J. and J.A. Hassell. 1984. Polyomavirus and simian virus 40 large T antigens bind to common DNA sequences. *J. Virol.* **49:** 925.

Pomerantz, B.J., C.R. Mueller, and J.A. Hassell. 1983. Polyomavirus large T antigen binds independently to multiple, unique regions on the viral genome. *J. Virol.* **47:** 600.

Tyndall, C., G. LaMantia, C.M. Thacher, J. Favaloro, and R. Kamen. 1981. A region of the polyoma virus genome between the replication origin and late protein coding sequences is required in *cis* for both early gene expression and viral DNA replication. *Nucleic Acids Res.* **9:** 6231.

Promoter Specificity of the SV40 Enhancer: Activation by T Antigen

P. Robbins and M. Botchan

Department of Molecular Biology, University of California
Berkeley, California 94720

Enhancers are *cis*-acting DNA elements essential for the efficient transcription of many viral and cellular genes. Enhancers differ from other *cis*-acting regulatory sequences in that they can function in an orientation- and position-independent manner, having been found 5' (Gruss et al. 1981; Fromm and Berg 1983; de Villiers and Schaffner 1981) and 3' (Lusky et al. 1983) to the transcription unit, within an intron (Banerji et al. 1983; Gillies et al. 1983; Queen and Baltimore 1983) as well as within the coding sequence itself (Osborne et al. 1984). Presumably, other *cis*-acting DNA sequences control the polarity of transcription. Enhancers also appear to display a cell-type specificity (de Villiers et al. 1982; Laimins et al. 1982; Lusky et al. 1983; Kriegler and Botchan 1983), most likely reflecting a requirement for specific cellular *trans*-acting factors for activity (Schöler and Gruss 1984). Enhancer function and thus gene expression may be regulated by the availability or activity of these specific factors. However, the mechanism by which an enhancer augments transcription from a given promoter is still a matter of conjecture. Furthermore, it is not yet clear if enhancers can act nonspecifically to augment expression from any linked promoter or if a given enhancer displays specificity for certain types of promoters.

To explore the last point, we have examined the interactions between the SV40 enhancer and two promoters, the SV40 early promoter and the Herpes simplex virus (HSV) thymidine kinase (*tk*) promoter, in several different cell types. We will describe experiments that show that the SV40 enhancer has no effect on the level of RNA expression from the *tk* gene, but that in the presence of the *trans* action of the SV40 T antigen, an interaction between enhancer and gene is detected. These observations lead to the conclusion that different promoters can be independently regulated by their specific enhancer interaction. Furthermore, the experiments suggest

41

that the apparent cell-type specificity of an enhancer may actually reflect the cell specificity of an enhancer-promoter pair.

RESULTS
Different Effects of the SV40 Enhancer on the HSV *tk* and SV40 Early Promoters

We have utilized two different sorts of constructions. In the first type the SV40 enhancer-promoter was fused to the body of the *tk* gene deleted for its own promoter. In these constructions the SV40 promoter drives *tk* gene expression. In the second set of constructions, we have linked the SV40 enhancer to the *tk* promoter–*tk* structural gene (see legend to Fig. 1 for a brief description). In all of the tested constructions, a small deletion at the *Bgl*I site functionally inactivates the SV40 origin of replication. By using the same structural gene as a marker for expression from the two promoters, we are able to contrast effects of the SV40 enhancer on promoter activity in different cell backgrounds. We initially examined the effects of the SV40 enhancer on these two promoters in transient assays in CV-1 and HeLa cell lines. Surprisingly, we did not detect augmentation of transcription from the *tk* promoter by the SV40 enhancer; i.e., in the second set of constructions the basal level of transcription from the *tk* gene was unaffected by the enhancer. In contrast, the presence of the SV40 enhancer strongly stimulated the level of expression from the SV40 early promoter in the fusion gene constructs. These observations suggest that the SV40 enhancer is unable to interact efficiently with the *tk* promoter to stimulate expression. However, when we examined the effect of the enhancer on the *tk* promoter in COS-7 and 293 cell lines, we detected an enhancer-dependent stimulation of *tk* transcription of 15-fold and 25-fold, respectively. SV40 early promoter expression in COS-7 cells still showed a strong requirement for the enhancer, whereas in 293 cells the requirement for the enhancer was alleviated, as has been previously noted (Green et al. 1983). These results are summarized in Table 1.

T Antigen Can Stimulate Enhancer-dependent *tk* Transcription in CV-1 Cells

The stimulation of enhancer-dependent *tk* transcription in COS-7 cells could be due to the presence of T antigen or to an unknown variable between CV-1 and COS-7 cells. To show that viral proteins are involved in the stimulation, we cotransfected the SV40-containing plasmid pMK16 SV40 6-1, which is ori-minus but expresses wild-type T antigen (Gluzman et al. 1980), along with either the enhan-

Table 1

Summary of the Effects of the SV40 Enhancer on the *tk* and SV40 Early Promoters in Different Cell Backgrounds

Promoter	Enhancer	Cell backgrounds			
		CV-1	COS-7	HeLa	293
tk	−	1	1	1	2–3
tk	+	1–2	20	1–2	50
SV40 early	−	1	1	1	40
SV40 early	+	50	50	50	80

Enhancer-minus transcription for both the *tk* and SV40 promoters in CV-1 and HeLa cells was assigned the relative value of 1. The table reflects a comparison of relative expression levels between CV-1, COS-7, HeLa, and 293 cell lines for the *tk* and SV40 early promoters; however, no direct comparison can be made between *tk* and SV40 early promoter expression since two different 5'-labeled probes were used to quantitate transcription.

cer-plus or enhancer-minus *tk* plasmids into CV-1 cells. We detected a 5-fold to 7-fold stimulation of enhancer-dependent *tk* transcription by T antigen (Fig. 1). Furthermore, by using a temperature-sensitive T antigen in cotransfection experiments, we have shown that the stimulation of enhancer-dependent *tk* transcription is at least in part due to the effects of the large T antigen. It is important to note that we have observed the T-antigen effects in the absence of the T-antigen-binding sites. Moreover, the effect observed in the presence of T antigen can also be detected in 293 cells, a human cell line that expresses E1A and E1B of adenovirus. These observations suggest that the effects of T antigen are independent of direct T-antigen binding to the SV40 origin and thus may be indirect. We are currently examining the effects of growth conditions and growth factors on SV40 enhancer-dependent *tk* expression to determine if the observed effect of T antigen on enhancer activity reflects a component of gene regulation in normal cells.

DISCUSSION
Enhancer-Promoter Specificity
The results presented above demonstrate that the interaction between the SV40 enhancer and a promoter in *cis* can be induced by the activity of SV40 T antigen in *trans*. The induction is clearly one that involves an enhancer interaction with the promoter, since the basal level of *tk* expression was *not* affected by T antigen in *trans* in the absence of a *cis*-acting SV40 enhancer (Fig. 1). This observation implies that the specific effects of the enhancer probably involve protein factors in addition to those involved in basal level transcrip-

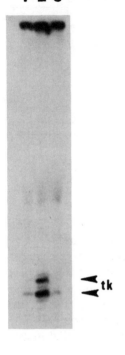

− + + T-antigen

+ + − Enhancer

1 2 3

◄ tk

Figure 1 Effect of T antigen on SV40 enhancer-dependent transcription in CV-1 cells. An SV40-containing plasmid, pMK16-SV40 6-1 (Gluzman et al. 1980), was cotransfected with an SV40 enhancer-plus (lane *2*) or an enhancer-minus (lane *3*) *tk* construction. As a control, a plasmid containing just the SV40 enhancer was cotransfected with the SV40 enhancer-plus *tk* plasmid (lane *1*). The RNA was isolated 48 hr posttransfection and analyzed by 5′ S1 mapping. A 131-base *Eco*RI-*Bgl*II 5′-labeled probe spanning the two *tk* RNA 5′ start sites was used, resulting in protected fragments of 54 and 56 bases in length (→). The enhancer-minus construct contains the 3.4-kb *Bam*HI *tk* fragment cloned into the *Bam*HI site of pML2. The enhancer-plus construct contains the SV40 *Hind*III-C fragment from position 1046 to 5171 cloned into the *Hind*III site of pMLTK. The results shown are with the early promoter of SV40 in the opposite orientation as the *tk* promoter. However, similar results have also been obtained with an SV40 fragment (position 117 to 273) containing only the enhancer element.

tion. The surprising feature of this induction is that it is clearly gene specific. It has been widely observed that the SV40 promoter can be stimulated by the *cis*-acting effects of its enhancer even in the absence of T antigen. Accordingly, we have found that when the SV40 early promoter is fused to the *tk* gene, thereby replacing the *tk* promoter, early gene expression is affected by the enhancer in CV-1, HeLa, and COS-7 cells. In contrast, *tk* gene expression mediated by the SV40 enhancer is only markedly affected in CV-1 cells in the presence of T antigen. These findings suggest that different balances of trancription factors are required for different pairs of promoters and enhancers. Furthermore, in assessing the cell specificity of enhancer-like sequences, it is clear that the nature of the promoter in question is a central factor. Thus it is not necessarily true that an enhancer has cell specificity, but that it is the particular enhancer-promoter interaction that shows cell-specific responses.

A Model for Enhancer-Promoter Interaction

We propose two models, one quantitative and the other qualitative to explain the observed differences between the enhancer effects on the HSV *tk* promoter and those on the SV40 early promoter. The quantitative model proposes that the requirement for enhancer factors for *tk* expression is greater than that for SV40 early promoter expression; the SV40 enhancer-promoter interaction is more efficient. Therefore, the SV40 enhancer is functioning in an analagous manner for each promoter, using the same enhancer-specific factors that interact with similar promoter-proximal transcription factors. The inability of the SV40 enhancer to stimulate the SV40 early promoter in 293 cells may reflect the bypass of enhancer function by stimulation of new, promoter-proximal, specific transcription factors that are not involved in expression from the *tk* promoter.

The qualitative model proposes that the SV40 enhancer is functioning with a slightly different set of transcription factors for the different promoters. In Figure 2 we illustrate two hypothetical ways in which this induction may act. At the top of the figure, a protein or complex of proteins is interacting with the SV40 enhancer. This enhancer complex makes an ineffective interaction with the transcription complex at the *tk* promoter, which thus leads to no augmentation of *tk* transcription. After induction by T antigen, new, specific enhancer factors are present (Fig. 2, B) that can now make a positive interaction with the transcriptional machinery at the *tk* promoter. However, since this requires that the same enhancer element bind different factors, another possibility would be that a new set of factors that interact with the upstream promoter sequences are induced (Fig. 2, A) that can now make a positive interaction with the enhancer complex.

An equally plausible model would be that a type of protein is induced that is not a DNA-binding protein but rather complexes to either the promoter-proximal or enhancer protein. Such tertiary complexes between enhancers and promoters would clearly provide multiple levels for regulation and could readily explain the range of effects seen in different constructions.

The model we present in Figure 2 is in some ways equivalent to the entry-site model for enhancers in that the protein interacting with the enhancer participates in the initiation of transcription event. However, the protein makes contact with the promoter through the bending of the DNA to interact with the closest promoter rather than by migration along the DNA. This bending model would account readily for the way in which an apparently asym-

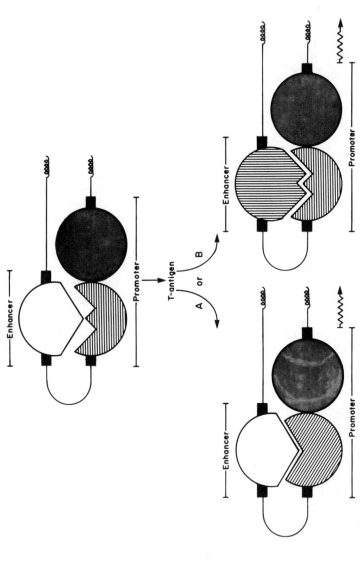

Figure 2 Two hypothetical models to explain the induction of an enhancer-promoter interaction. For description, see Discussion.

metric DNA sequence (Weiher and Botchan 1984; M. Zenke et al., pers. comm.) could interact in both orientations with respect to the direction prescribed by the promoter. We are currently determining the sequences within the *tk* promoter through which the SV40 enhancer mediates its stimulatory effect on transcription.

REFERENCES

Banerji, J., L. Olson, and W. Schaffner. 1983. A lymphocyte-specific cellular enhancer is located downstream of the joining region in immunoglobulin heavy chain genes. *Cell* **33:** 729.

de Villiers, J. and W. Schaffner. 1981. A small segment of polyoma enhances the expression of the cloned β-globin gene over a distance of 1400 base pairs. *Nucleic Acids Res.* **9:** 6251.

de Villiers, J., L. Alson, C. Tyndall, and W. Schaffner. 1982. Transcriptional "enhancers" from SV40 and polyoma show a cell type preference. *Nucleic Acids Res.* **10:** 7965.

Fromm, M. and P. Berg. 1983. Deletion mapping of DNA regions required for SV40 early gene expression *in vivo*. *J. Mol. Appl. Genet.* **1:** 457.

Gillies, S.D., S.L. Morrison, V.T. Oi, and S. Tonegawa. 1983. A tissue specific enhancer element is located in the major intron of a rearranged immunoglobin heavy chain gene. *Cell* **33:** 717.

Gluzman, Y., J.F. Sambrook, and R.J. Frisque. 1980. Expression of early genes of origin-defective mutants of simian virus 40. *Proc. Natl. Acad. Sci.* **77:** 3898.

Green, M.R., R.H. Treisman, and T. Maniatis. 1983. Activation of globin gene transcription in transient assays by *cis* and *trans*-activating functions. *Cell* 35: 137.

Gruss, P., R. Dhar, and G. Khoury. 1981. Simian virus repeats as an element of the early promoter. *Proc. Natl. Acad. Sci.* **78:** 993.

Kriegler, M. and M. Botchan. 1983. Enhanced transformation by a simian virus 40 recombinant virus containing a Harvey murine sarcoma virus long terminal repeat. *Mol. Cell. Biol.* **3:** 325.

Laimins, L.A., G. Khoury, C. Gorman, B. Howard, and P. Gruss. 1982. Host specific activation of transcription by tandem repeats form simian virus 40 and a Moloney murine leukemia virus. *Proc. Natl. Acad. Sci.* **79:** 6453.

Lusky, M., L. Berg, H. Weiher, and M. Botchan. 1983. Bovine papilloma virus contains an activator of gene expression at the distal end of the early transcription unit. *Mol. Cell. Biol.* **3:** 1108.

Osborne, T.F., D.N. Arvidson, E.S. Tyau, M. Dunsworth-Browne, and A. Berk. 1984. Transcriptional control region within the protein-coding portion of adenovirus E1A genes. *Mol. Cell. Biol.* **4:** 1293.

Queen, C. and D. Baltimore. 1983. Immunoglobulin gene transcription is activated by downstream sequence elements. *Cell* **33:** 741.

Schöler, H.R. and P. Gruss. 1984. Specific interaction between enhancer-containing molecules and cellular components. *Cell* **36:** 403.

Weiher, H. and M. Botchan. 1984. An enhancer sequence for bovine papilloma virus DNA consists of two essential regions. *Nulceic Acids Res.* **12:** 2901.

Adenovirus Enhancer Elements

P. Hearing* and T. Shenk †

*Department of Microbiology, Health Sciences Center
State University of New York at Stony Brook
Stony Brook, New York 11794

†Department of Molecular Biology, Princeton University
Princeton, New Jersey 08540

We previously identified an enhancer region located in the left end of the adenovirus type-5 (Ad5) genome that positively regulates transcription of region E1A in a location and orientation-independent fashion (Hearing and Shenk 1983a). This region also potentiates transcription from heterologous promoters in transient and long-term expression assays (Hearing and Shenk 1983a). Here we describe a detailed mutational analysis of this region. Our results define two separate, functional enhancer elements located in the adenovirus enhancer region. The first element is repeated, resembles enhancer core sequences from several other eukaryotic enhancers, and specifically regulates transcription from region E1A. The second element is located between these repeated sequences and regulates transcription from the entire adenovirus chromosome.

Construction and Analysis of Viral Mutants

Deletion mutations were initially constructed in a recombinant plasmid, pE1A-WT, that contains the left 1339 bp of Ad5 cloned in pBR322 (Hearing and Shenk 1983a). Small, random deletion mutations were introduced through the enhancer region (Ad5 nucleotides 194–353), using a D-loop mutagenesis protocol (Hearing and Shenk 1983b). The exact location of each deletion was determined by nucleotide sequence analysis and is shown in Figure 1. The region that we previously defined as an enhancer (Hearing and Shenk 1983a) is located between nucleotides 194 and 353. The mutations were then rebuilt into intact viral chromosomes, using the method of Stow (1981). Viruses were propagated in 293 cells (a permissive human cell line that contains and expresses Ad5 regions E1A and E1B; Graham et al. 1977) and subsequently used to infect HeLa cells, which express no adenovirus gene products, for analysis of early viral transcription.

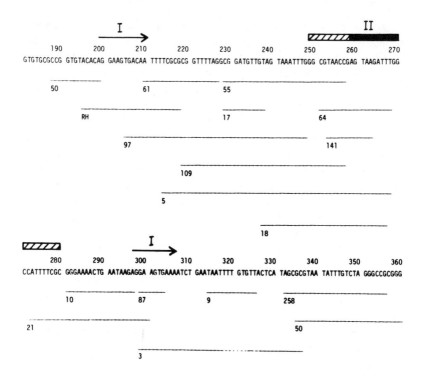

I CORE = AGGAAGTGAAA

II CORE = AGTAAGATTTG

Figure 1 Deletions constructed in the Ad5 enhancer region. The nucleotide sequence of the Ad5 enhancer region is presented. The numbers above the DNA sequence are relative to the left end of the Ad5 genome. Deletions were constructed using a D-loop mutagenesis procedure and are indicated by bars below the DNA sequence. Ad5 enhancer elements I and II are indicated above the nucleotide sequence. Core enhancer sequences are also presented.

The effects of these mutations on E1A transcription in vivo were analyzed within infected cells. HeLa cells were infected at a multiplicity of infection of 20, and total cytoplasmic RNA was isolated early (6 hr) after infection. Levels of E1A transcripts were analyzed by S1 nuclease analysis using a uniformly ^{32}P-labeled M13-E1A probe. Our results demonstrated that two functional elements within the enhancer region regulate transcription of region E1A.

The first is the repeated sequence (element I in Fig. 1, designated by arrows over the sequence), which we previously described, that contains a core sequence conserved in several other enhancer regions. Deletion of either the upstream or downstream copy of this element did not reduce E1A transcription in vivo, but deletion of both repeated elements (combined deletions RH and 87 in Fig. 1) reduced E1A transcription 4-fold to 5-fold. Deletions within region II in Figure 1 also resulted in reduced E1A transcription in vivo. Deletion of the core of this region (mutants 64 and 141) resulted in a 4-fold to 5-fold reduction in E1A transcription within infected cells. Mutants containing deletions approaching the core of element II (deletions 55, 109, and 21 in Fig. 1) also showed reduced levels of E1A transcripts, but the level of reduction was only 2-fold to 3-fold compared with the wild-type virus.

We previously reported that the Ad5 enhancer region augmented E1A transcription when located at the 3' end of the region E1A (+1275 relative to the cap site at +1; Hearing and Shenk 1983a). We wished to determine if elements I and II contained this enhancer property. Ad5 variants were constructed that contained a deletion of the entire enhancer region upstream of E1A and an insertion of the same region at the 3' end of the E1A transcription unit containing the wild-type sequence or carrying deletion RH/87 or deletion 64. E1A transcription was again analyzed at early times after infection of HeLa cells. Deletion of the entire enhancer region upstream of E1A reduced E1A transcription about 20-fold as previously described (Hearing and Shenk 1983a). Insertion of this region containing wild-type sequences at the 3' end of region E1A increased E1A transcription approximately 5-fold. Increased levels of E1A transcripts were also observed with variants carrying the enhancer region at the 3' end of region E1A carrying either deletion RH/87 or deletion 64. In both cases, however, levels of E1A transcripts were reduced slightly compared with the virus containing an intact enhancer at the 3' end of region E1A. Both elements I and II, therefore, function when inserted at the 3' end of region E1A, although less well than when located in their normal location.

We also analyzed the effects of deletion of elements I and II on transcription from the other regions expressed at early times after infection with adenovirus. The E1A gene products are required in *trans* to positively regulate transcription of the other early regions in the adenovirus chromosome (Berk et al. 1979; Jones and Shenk 1979). These experiments, therefore, were performed in 293 cells to provide wild-type levels of E1A gene products to mutants expressing reduced levels of E1A proteins due to reduced E1A transcription. Cytoplasmic RNA was isolated early after infection, and polyadenylated RNA was analyzed by Northern hybridization analysis using probes specific for Ad5 early regions E2, E3, and E4. Transcription from the early regions tested was at wild-type levels with the mutant carrying the RH/87 deletion as well as with an E1A coding-region mutant (used as a control for 293 cell E1A function). In contrast, transcription from each of the early regions was reduced with the mutant carrying deletion 64. E2 and E3 transcription were each reduced 4-fold to 5-fold and E4 transcription was reduced about 3-fold. This effect was also observed using viruses carrying larger deletions that overlap element II that also contain the Ad5 packaging sequences and enhancer region at the right end of the viral chromosome. Decreased early region transcription with mutants containing deletions of element II was not complemented in *cis*, therefore, when this region was placed at the right end of the Ad5 chromosome.

DISCUSSION

We have defined two separate, functional enhancer elements that regulate adenovirus transcription in vivo within infected cells. The first element (I) is repeated, resembles sequences in several other enhancer regions, and specifically regulates E1A transcription. A perfectly homologous sequence is found in the polyoma virus enhancer region and has been shown recently to be a major functional component of the polyoma enhancer (Herbomel et al. 1984; Hassell et al., this volume). A second enhancer element (II) is located between the repeated sequences and regulates transcription from the entire adenovirus chromosome. Detailed mutational analysis of several enhancer regions recently has shown that SV40 and polyoma enhancer regions each contain a complex set of transcriptional regulatory sequences that are often repeated and may be functionally redundant (Herr et al.; Hassell et al.; both this volume). In this respect the Ad5 enhancer region is similar since two functional enhancer elements have been identified, one of which is repeated (element I). Although the nucleotide sequences that constitute the Ad5 enhancer region have been defined, we know nothing about the functional role of these enhancers in the regulation of adenovirus transcription. We are currently analyzing the effect of the deletion of Ad5 elements I and II on the assembly of viral chromatin early after infection and on the association of viral DNA with the nuclear matrix.

Enhancer element II lies in close proximity to sequences required in *cis* for efficient packaging of adenovirus DNA into virions (P.

Hearing and T. Shenk, unpubl.). An argument, therefore, could be made that deletion of this region reduces adenovirus transcription by affecting an unpackaging event early after infection. Genetic data strongly argue against this possibility. The packaging sequences do not function when located at the 3' end of region E1A, whereas element II functions at this position. Conversely, the packaging sequences function perfectly well when located at the right end of the viral genome, whereas element II only functions when located at the left end. We have no clear explanation for the polarity observed with element II with respect to the ends of the viral chromosome. Perhaps a sequence at the right end poisons the function of element II. Alternatively, a functional equivalent of this region may already exist at the right end of the viral chromosome. Interestingly, a 12-bp sequence is located at the right end of the Ad5 genome (180–200 bp from the right terminus; Steenbergh and Sussenbach 1979) that shares an 11 of 12 nucleotide homology with the core of element II (solid bar above the sequence in Fig. 1). Element II, therefore, may be duplicated at each end of the viral chromosome. This possibility is currently being analyzed.

REFERENCES

Berk, A.J., F. Lee, T. Harrison, J. Williams, and P.A. Sharp. 1979. Pre-early adenovirus 5 gene product regulates synthesis of early viral messenger RNAs. *Cell* **17:** 935.

Graham, F.L., J. Smiley, W.C. Russell, and R. Nairu. 1977. Characteristics of a human cell line transformed by DNA from human adenovirus type 5. *J. Gen. Virol.* **36:** 59.

Hearing, P. and T. Shenk. 1983a. The adenovirus type 5 E1A transcriptional control region contains a duplicated enhancer element. *Cell* **33:** 695.

―――. 1983b. Functional analysis of the nucleotide sequence surrounding the cap site for adenovirus type 5 region E1A messenger RNAs. *J. Mol. Biol.* **167:** 809.

Herbomel, P., B. Bourachot, and M. Yaniv. 1984. Two distinct enhancers with different cell specificities coexist in the regulatory region of polyoma. *Cell* **39:** 653.

Jones, N. and T. Shenk. 1979. An adenovirus type 5 early gene function regulates expression of other early viral genes. *Proc. Natl. Acad. Sci.* **76:** 3665.

Steenbergh, P.H. and J.S. Sussenbach. 1979. The nucleotide sequence of the right-hand terminus of adenovirus type 5 DNA: Implications for the mechanism of DNA replication. *Gene* **6:** 307.

Stow, N.D. 1981. Cloning of a DNA fragment from the left-hand terminus of the adenovirus type 2 genome and its use in site-directed mutagenesis. *J. Virol.* **37:** 171.

Control of Viral Expression in Undifferentiated Teratocarcinoma Cells

C. Gorman,* P. Rigby, and D. Lane

Cancer Research Campaign
Eukaryotic Molecular Genetics Research Group
Department of Biochemistry, Imperial College of Science and Technology
London SW7 2AZ, England

Many viruses share the common property of failure to replicate or produce virions within undifferentiated teratocarcinoma cells. This phenomenon has been seen for both DNA and RNA viruses. It appears that the virus does enter the cell and can be found as integrated copies. Yet there is a block to the formation of virus. Although much of the DNA within the undifferentiated cells is highly methylated, the *de novo* methylation of the incoming viral genome is a delayed event, suggesting it is not the primary block to expression.

It has been suggested that regulatory sequences within the virus are nonfunctional in the undifferentiated cells. It is possible to isolate mutant polyoma virus that can function in these cells. Almost entirely, these mutants can now be classified as enhancer mutants (for review, see Silver et al. 1983). It has recently been suggested that the lack of expression of Moloney murine sarcoma virus (Mo-MSV) is also due to the lack of function of the enhancer (Linney et al. 1984).

Studies on Expression of SV40 T Antigen

We have studied the expression of viral promoters in teratocarcinoma cells. We used either an antibody staining assay to detect viral gene products or assayed promoter activity directly by use of the choramphenicol acetyltransferase (CAT) assay. Our studies of viral

---

Present address: Department of Molecular Biology, Genentech, Inc., 460 Point San Bruno Boulevard, South San Francisco, California 94080.

expression in teratocarcinoma cells began by repeating previous experiments that showed that no large T antigen is detected in the undifferentiated cells after viral infection. We used a high multiplicity of SV40 for infection. When cells were screened for production of large T antigen by antibody staining, only a very small proportion of cells, 0.01%, stained positively (Fig. 1A). We concurrently stained with SSEA-1, which is specific for the undifferentiated cells (Solter and Knowles 1978), and TROMA-1, an antibody against intermediate filaments (see Silver et al. 1983), which begin to appear in our cultures 24–30 hr after the addition of retinoic acid (RA). The expected pattern of dense SSEA-1 staining with fewer than 0.01% of the cells staining with TROMA-1 was obtained.

After DNA transfection of SV40 DNA, full-length large T antigen and small T antigen were produced. Both large T and small T antigens are made in up to 50% of the cells after transfection (Fig. 1B). The expression of large T antigen after transfection with the viral DNA is influenced by DNA concentration. When less than 1 μg of viral DNA was used for transfection (plus 9 μg of carrier), less than 10% of the undifferentiated cells expressed T antigen. The dependence on DNA concentration was not seen in transfections of differentiated cells. Furthermore, the same number of cells were positive after a 72-hr treatment of RA, regardless of the varying amounts of DNA used. Control cells stained with the appropriate markers clearly showed that the cells that express T antigen are undifferentiated.

Previous studies have suggested that some factor necessary for expression of SV40 was lacking in the undifferentiated cells. However, from our results showing 50% of the cells capable of expressing T antigen after transfection, we conclude that at high DNA concentration the SV40 early region genes are transcribed. These results suggest that, indeed, all the factors required for SV40 expression are present in the undifferentiated cells.

Figure 1 *(see facing page)* (A) Undifferentiated F9 cells were infected with SV40 virus. Forty-eight hr after infection, cells were fixed for antibody staining. Several monoclonal antibodies for large T antigen were used, but no positive cells were seen. In control plates of NIH-3T3 cells, up to 50% showed positive large-T staining after infection. (B) Undifferentiated F9 cells were transfected with 25 μg of SV40 DNA per 60-mm dish. Forty-eight hr later cells were fixed and assayed for large T antigen by immunochemical staining. Up to 50% of the cells stain positive for large T antigen after transfection.

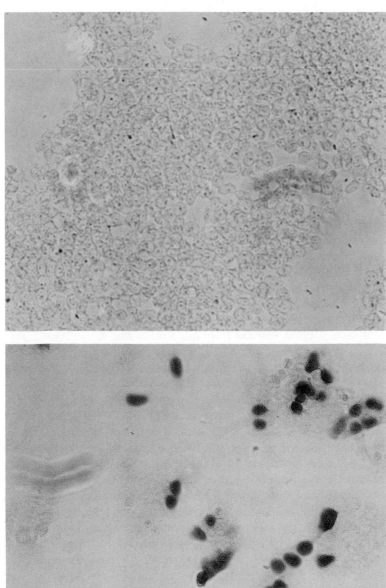

Figure 1 *See facing page for legend.*

Study of Promoter Activity

To study promoter/enhancer function in detail, we used vectors expressing CAT (Gorman et al. 1982). The Rous sarcoma virus (RSV) long terminal repeat (LTR) functions efficiently in the undifferentiated cells. Transcription from the SV40 early region does occur, as can be seen by CAT assays. A detailed analysis reveals that this expression is enhancer independent since we see no difference in the level of expression of CAT when the SV40 enhancer is removed. In differentiated murine cells, this deletion decreases the activity of the SV40 early region approximately 10-fold (Fig. 2).

We also studied expression of the Mo-MSV LTR. The entire LTR directs very little or no CAT synthesis. However, further data suggest that a cellular factor could interefere with the enhancer region of the LTR. This is illustrated by the fact that a deletion of the Mo-MSV 5' LTR, including the enhancer, can restore transcription from the LTR cap site. This is seen with the construct pLTR10 (Laimins et al. 1984). Another deletion vector, pLTR0 (Laimins et al. 1984), shows no expression activity. Linney et al. (1984) have shown that a deletion of the Mo-MSV LTR identical to the deletion in pLTR0 cap is nonfunctional in teratocarcinoma cells, undifferentiated and differentiated. We stress that from our studies this construct is incapable of transcription in any cell type. As is illustrated in Figure 2, once the teratocarcinoma cells have been treated with RA, complete enhancer function is restored to the LTR.

The results with the Mo-MSV clearly suggest that regulation of enhancers could be due to the presence of a factor that can interfere with enhancer function. This idea is supported by the enhancer-independent expression of the SV40 promoter. These results are suggestive of the possibility of the presence of a *trans*-acting factor present in the undifferentiated cells, as seen in 293 cells. These cells express a *trans*-acting viral protein (Treisman et al. 1983), E1A, that can influence expression of viral enhancers (Borrelli et al. 1984). From our data it seems likely that a *trans*-acting factor is present in the undifferentiated cell that can prevent viral expression or perhaps replication, since both of these functions require enhancer activity. The undifferentiated cells do contain an E1A-like factor (Imperiale et al. 1984). The fact that this factor decreases with differentiation at a time concurrent with the expression of viral components suggests that the presence of such a factor in the undifferentiated cell could prevent viral expression. Such a factor, capable of repressing viral expression, may also have a primary role in influencing the expression pattern of cellular genes induced by differentiation.

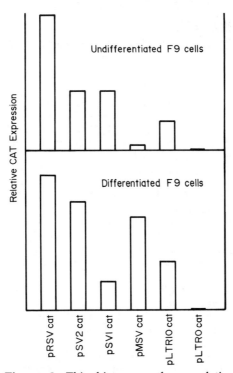

Figure 2 This histogram shows relative promoter strengths of assayed undifferentiated and differentiated F9 cells. The transfection efficiency of each cell type (in undifferentiated vs. differentiated) is not the same; therefore comparisons are made only within one cell type. (pRSVcat) The entire RSV LTR is used as a promoter. (pSV2cat) The entire SV40 early region, including the enhancer, is used in this plasmid. (pSV1cat) This plasmid uses the start site of SV40 early region, plus the 21-bp GC-rich repeats, but has the 72-bp enhancers deleted. (pMSVcat) This plasmid contains the entire Mo-MSV LTR as a promoter. (pLTR10) This construct uses the Mo-MSV LTR but has deleted the enhancer and upstream region; it contains the GC-rich repeats. (pLTR0) This plasmid has both the enhancer and GC-rich repeats deleted; it is transcriptionally inactive in all cell types tested. Sixty-mm dishes were transfected with 2.5 μg of test plasmid and 2.5 μg of carrier. Forty-eight hr after transfection, cells were harvested for the CAT assay (Gorman et al. 1982).

ACKNOWLEDGMENT

C.G. was a recipient of a NATO fellowship and an EMBO long-term fellowship.

REFERENCES

Borrelli, E., R. Hen, and P. Chambon. 1984. Adenovirus-2 E1a products repress enhancer induced stimulation of transcription. *Nature* **312:** 608.

Gorman, C.M., L.F. Moffat, and B.H. Howard. 1982. Recombinant genomes which express chloramphenicol acetyl transferase in mammalian cells. *Mol. Cell. Biol.* **2:** 1044.

Imperiale, M.J., H.T. Kao, L. Feldman, J. Nevins, and S.D. Strickland. 1984. Common control of heat shock gene and early adenovirus genes: Evidence for a cellular E1a-like activity. *Mol. Cell. Biol.* **4:** 867.

Laimins, L.A., P. Gruss, R. Pozzatti, and G. Khoury. 1984. Characterization of enhancer elements in the long terminal repeat of Moloney murine sarcoma virus. *J. Virol.* **49:** 183.

Linney, E., B. Davis, J. Overhauser, E. Chao, and H. Fan. 1984. Nonfunction of a regulatory sequence in F9 embryonal carcinoma cells. *Nature* **308:** 470.

Silver, L., G. Martin, and S. Strickland, eds. 1983. *Teratocarcinoma stem cells. Cold Spring Harbor Conf. Cell Proliferation*, vol. 10.

Solter, D. and B. Knowles. 1978. Monoclonal antibody defining a stage-specific mouse embryonic antigen (SSEA-1). *Proc. Natl. Acad. Sci.* **75:** 5565.

Treisman, R., M. Green, and T. Maniatis. 1983. *Cis* and *trans* activation of globin gene transcription in transient assays. *Proc. Natl. Acad. Sci.* **80:** 7428.

Constitutive and Inducible Enhancer Elements

E. Serfling, M. Jasin, and W. Schaffner

Institut für Molekularbiologie II der Universität Zürich
Hönggerberg, CH-8093 Zürich, Switzerland

In higher eukaryotes, gene transcription is regulated via *cis*-acting DNA motifs that apparently interact with specific protein factors. Many, but not all, of these sequences are located in the promoter region just upstream of the transcription initiation site. These regulatory sequences are quite diverse in their function, as well as in their structure, acting in the general stimulation of transcription, tissue-specific gene expression, or the induction (or repression) of transcription by the action of specific agents.

We and others had previously shown that some of these regulatory sequences have an unexpected property that has been termed the enhancer effect (Banerji et al. 1981; Moreau et al. 1981; Fromm and Berg 1983). The enhancer sequences upstream of the early transcription unit of SV40 were found to stimulate transcription of a heterologous gene by more than two orders of magnitude in either orientation and over distances of more than 3000 bp, even from a position downstream of the gene (Banerji et al. 1981). Although upstream promoter components had been demonstrated to be flexible with regard to orientation, and to some extent, distance from the cap site, the downstream and long-range activation were novel properties. Enhancers have since been described for a variety of viral and cellular genes (for review, see Gluzman and Shenk 1983; Picard 1985). Typically, an enhancer is about 200 bp long and harbors several conserved sequence motifs (Banerji et al. 1983; Hearing and Shenk 1983; Weiher et al. 1983; Herr and Gluzman 1985) that are presumably protein-binding sites (Schöler and Gruss 1984; Seguin et al. 1984; Wildeman et al. 1984).

We have detected an inducible enhancer within the upstream region of the mouse metallothionein-I (MT-I) gene. In transient expression assays, the metallothionein enhancer, in particular if present in several tandem copies, stimulates the expression of a linked β-globin gene over a large distance from a 3' position. Analysis of the sequences within this enhancer indicates that it is composed of several regulatory and constitutive sequences motifs, some of which are also present within the upstream regions of other genes. These results indicate that "enhancers" and "promoters" overlap physically and functionally and that upstream promoter elements can show enhancer activity when detached from the most proximal promoter components such as the TATA box (see also Weber and Schaffner 1985).

The Enhancer of the Mouse MT-I Gene
The metallothionein genes of mouse and man belong to one of the best-studied gene families. Transcriptional induction of these genes is elicited by several stimuli, notably heavy-metal ions. To determine if the upstream sequences of the mouse MT-I gene harbor an enhancer, we have used the SV40 "enhancer trap" approach (Weber et al. 1984). Noninfectious, linear SV40 DNA lacking the enhancer was cotransfected with sonicated MT-I DNA into monkey CV-1 cells that were incubated in the presence of zinc and cadmium ions. Infectious virus that had integrated MT-I sequences was recovered. Two independent, variant SV40 viruses were found to have integrated similar upstream segments of the MT-I gene from approximately -190 to -60 in the opposite orientation with respect to the SV40 early promoter (Fig. 1A). Recovery of SV40–MT-I recombinant viruses was dependent upon heavy metal treatment of the CV-1 cells during the incubation period.

To determine if the MT-I sequences are able to activate gene transcription from a downstream position, they were cloned 3' of a rabbit β-globin test gene. After transfection into human HeLa cells that had been treated with zinc and cadmium ions, a weak stimulation of β-globin transcription was detected. However, these enhancer sequences appear to exert a cumulative effect since considerably stronger transcription was detected with two or four tandem copies of the MT-I enhancer inserted downstream of the β-globin gene (Fig. 1B). The MT-I segment, therefore, confers metal regulation, indicating that the recovered MT-I sequences constitute an inducible enhancer.

Inducible and Constitutive Enhancer Elements
The upstream regions of metallothionein genes are composed of several different sequence motifs that are involved in transcriptional regulation. At about position -50 from the cap site, metallothionein promoters contain two adjacent motifs in an inverted configuration that have been demonstrated, at least for the mouse MT-I and human MT-IIA genes, to represent metal-inducible elements

(Fig. 1C). Recent evidence suggests that there are additional metal-responsive elements in the upstream regions of these genes. The additional elements share distinct sequence homologies to the proximal −50 motifs (Carter et al. 1984; Karin et al. 1984; Stuart et al. 1984).

The hormonal induction by glucocorticoids of one of the metallothionein genes, the human MT-IIA gene, is mediated by a DNA segment located at position −268 to −237, which does not overlap with the metal-responsive elements (Karin et al. 1984). Yet another DNA segment, mapped between −350 and −185 for the mouse MT-I is responsible for induction by bacterial lipopolysaccharides (Durnam et al. 1984).

In addition to these regulatory sequence elements, there seem to be interspersed constitutive elements. There are two CCGCCC elements present in an inverted configuration (Fig. 1C). Such CCGCCC motifs are contained within other cellular promoters as well as within the promoters of several viral genes (e.g., the SV40 early transcription unit [see Jones and Tjian, this volume]; the Herpes virus thymidine kinase gene [see McKnight et al., this volume]; and the major immediate early gene of the human cytomegalovirus [Boshart et al. 1985]). These elements are recognized by the positively acting transcription factor Sp1 (see Jones and Tjian, this volume). Besides the CCGCCC repeats, there are several short stretches of alternating purine/pyrimidine that may contribute to efficient transcription (Hamada et al. 1984; Karin et al. 1984; Herr and Gluzman 1985).

Thus, the upstream regions of metallothionein genes harbor a variety of regulatory elements that are intermingled with constitutive elements. The enhancer we have isolated from the MT-I gene contains many of these elements (Fig. 1C). The metal-responsive elements can apparently function independently from each other because the one closest to the TATA box is missing in the MT-I sequence recovered in our enhancer trap (Fig. 1C; see also Palmiter et al., this volume). Another striking example of an enhancer that is composed of several types of conserved sequence motifs is the one of human cytomegalovirus. In this very strong enhancer the information is highly redundant, since several subsegments can function independently (Boshart et al. 1985).

CONCLUSIONS

Our current view is that enhancers and upstream promoter elements are composed of a modular arrangement of short sequence motifs, each with a specific function in conferring inducibility, tis-

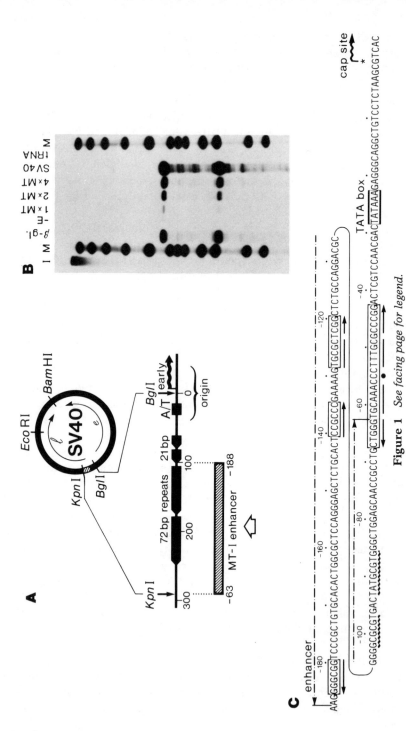

Figure 1 *See facing page for legend.*

Figure 1 (*see facing page*) The metallothionein enhancer. (*A*) The SV40 enhancer trap experiment. Enhancerless SV40 DNA (deletion from nucleotide 99 to the *Kpn*I site at nucleotide 294, according to the SV40 numbering system) was cotransfected with sonicated DNA from the mouse MT-I gene into CV-1 cells treated with 200 μM Zn^{++} and 2 μM Cd^{++}. One of the recombinant SV40-MT-I clones was found to have incorporated MT-I upstream sequences from -188 to -63. (*B*) Downstream activation of the β-globin gene by the MT-I enhancer. Sp6 mapping was performed on RNA isolated from HeLa cells after transfection of the recombinant β-globin clones. $1 \times$MT, $2 \times$MT, and $4 \times$MT are from recombinant clones containing one, two, and four copies, respectively, of the MT-I enhancer; ($-$E) no enhancer; (SV40) SV40 enhancer; (β-gl.) β-globin mRNA positive control; (M) marker; (I) input hybridization probe; (tRNA) tRNA as input RNA. The upper band corresponds to a protected fragment from a portion of the second β-globin exon; the lower band, from the first β-globin exon. (*C*) Sequence of the upstream region of the MT-I gene. Metal-responsive elements and the inverted CCGCCC sequences are boxed. Wavy lines indicate alternating purine/pyrimidine stretches. (--------) The enhancer.

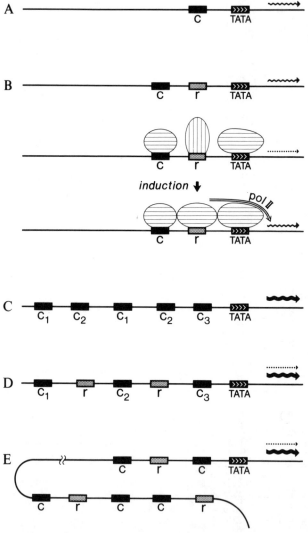

Figure 2 *See facing page for legend.*

sue specificity, or a general enhancement of transcription (Fig. 2). It is also possible that the difference in structure between strongly and weakly transcribed promoters is primarily of a quantitative rather than a qualitative nature. That is, the many sequence motifs used to build an enhancer of a strongly expressed gene could be present in only a few copies in the promoter of a weakly expressed gene (Fig. 2; see also Serfling et al. 1985). Further modulation could be achieved by varying the precise sequence of these motifs, resulting in differential binding of the motif's specific protein factor. The fact that regulatory DNA sequences of higher eukaryotes are composed of a variety of sequence motifs that are flexible with respect to orientation and distance from the site of transcription initiation (the "enhancer effect") would facilitate the development of complex regulatory networks. The development of such networks probably formed the basis for the evolution of multicellular organisms.

Figure 2 (*see facing page*) Simplified model of eukaryotic transcription control. (*A*) Promoters of weakly expressed genes contain only a few upstream DNA elements for constitutive (c) expression. (*B*) Promoters regulated at the level of cell differentiation or by induction with environmental stimuli harbor not only constitutive but also regulatory (r) motifs. A schematic view of regulation of transcription by the interaction of several factors, namely a TATA-box-binding protein, a constitutive factor (e.g., Sp1), and a regulatory factor (e.g., a steroid hormone receptor) is shown underneath. In the uninduced state, the latter factor could be absent, could be present but not bind, or, as shown, could bind in a conformation that does not stimulate but rather represses transcription. Induction presumably involves "bridging" of constitutive factors by a regulatory factor. (*C*) Promoters of strongly expressed constitutive genes. DNA motifs are arbitrarily numbered to illustrate that several types of elements can be intermingled and repeated. Although certain DNA motifs may occur preferentially in one orientation and close to the TATA box, and others further upstream, the difference between *A* and *C* may well be of a quantitative rather than a qualitative nature. (Motifs C_1-C_2-C_1-C_2 would exert an enhancer effect if detached and linked to a test gene.) (*D*) Promoters of strongly expressed, regulated genes. This example is an extension of situation *B*, with several regulatory motifs interspersed at strategic positions with constitutive elements. The number and spatial arrangement of these motifs probably determines the basal vs. induced level of transcription. (*E*) Remote control by an enhancer. The sequences just upstream of the TATA box are not sufficient to result in the induction of a high level of transcription by themselves, but they respond to additional elements far away, presumably mediated by protein-protein contacts of the factors involved. These remote elements may be located either within the trancriptional unit (as in immunoglobulin genes) or far upstream or downstream.

REFERENCES

Banerji, J., L. Olson, and W. Schaffner. 1983. A lymphocyte-specific cellular enhancer is located downstream of the joining region in immunoglobulin heavy chain genes. *Cell* **33:** 729.

Banerji, J., S. Rusconi, and W. Schaffner. 1981. Expression of a globin gene is enhanced by remote SV40 DNA sequences. *Cell* **27:** 299.

Boshart, M., F. Weber, G. Jahn, K. Dorsch-Häsler, B. Fleckenstein, and W. Schaffner. 1985. A very strong enhancer is located upstream of an immediate early gene of human cytomegalovirus. *Cell* **41:** 521.

Carter, A.D., B.K. Felber, M. Walling, M.-F. Jubier, C.J. Schmidt, and D.H. Hamer. 1984. Duplicated heavy metal control sequences of the mouse metallothionein-I gene. *Proc. Natl. Acad. Sci.* **81:** 7392.

Durnam, D.M., J.S. Hoffman, C.J. Quaife, E.P. Benditt, H.Y. Chen, R.L. Brinster, and R.D. Palmiter. 1984. Induction of mouse metallothionein-I mRNA by bacterial endotoxin is independent of metals and glucocorticoid hormones. *Proc. Natl. Acad. Sci.* **81:** 1053.

Fromm, M. and P. Berg. 1983. Simian virus 40 early- and late-region promoter functions are enhanced by the 72 base pair repeat inserted at distant locations and inverted orientations. *Mol. Cell. Biol.* **3:** 991.

Gluzman, Y. and T. Shenk, eds. 1983. *Current communications in molecular biology: Enhancers and eukaryotic gene expression.* Cold Spring Harbor Laboratory, Cold Spring Harbor, New York.

Hamada, H., M. Seidman, B.H. Howard, and C.M. Gorman. 1984. Enhanced gene expression by the poly(dT-dG)·poly(dC-dA) sequence. *Mol. Cell. Biol.* **4:** 2622.

Hearing, P. and T. Shenk. 1983. The adenovirus type 5 E1a transcriptional control region contains a duplicated enhancer element. *Cell* **33:** 695.

Herr, W. and Y. Gluzman. 1985. Duplications of a mutated SV40 enhancer restores its activity. *Nature* **313:** 711.

Karin, M., A. Haslinger, H. Holtgreve, R.I. Richards, P. Krauter, H.M. Westphal, and M. Beato. 1984. Characterization of DNA sequences through which cadmium and glucocorticoid hormones induce human metallothionein-II$_A$ gene. *Nature* **308:** 513.

Moreau, P., R. Hen, B. Wasylyk, R. Everett, M.P. Gaub, and P. Chambon. 1981. The SV40 72 base repair repeat has a striking effect on gene expression both in SV40 and other chimeric recombinants. *Nucleic Acids Res.* **9:** 6047.

Picard, D. 1985. Viral and cellular transcription enhancers. In *Oxford surveys on eukaryotic genes* (ed. N. Maclean), vol. 2. Oxford University Press, England. (In press.)

Schöler, H.R. and P. Gruss. 1984. Specific interaction between enhancer-containing molecules and cellular components. *Cell* **36:** 403.

Seguin, C., B.K. Felber, A.D. Carter, and D.H. Hamer. 1984. Competition for cellular factors that activate metallothionein gene transcription. *Nature* **312:** 781.

Serfling, E., M. Jasin, and W. Schaffner. 1985. Enhancers and eukaryotic gene transcription. *Trends Genet.* (in press).

Stuart, G.W., P.F. Searle, H.Y. Chen, R.L. Brinster, and R.D. Palmiter. 1984. A 12-base-pair DNA motif that is repeated several times in metallothionein gene promoters confers metal regulation to a heterologous gene. *Proc. Natl. Acad. Sci.* **81:** 7318.

Weber, F. and W. Schaffner. 1985. Enhancer activity correlates with the oncogenic potential of avian retroviruses. *EMBO J.* **4:** 949.

Weber, F., J. de Villiers, and W. Schaffner. 1984. An SV40 "enhancer trap" incorporates exogenous enhancers or generates enhancers from its own sequences. *Cell* **36:** 983.

Weiher, H., M. König, and P. Gruss. 1983. Multiple point mutations affecting the simian virus 40 enhancer. *Science* **219:** 626.

Wildeman, A.G., P. Sasonne-Corsi, T. Grundström, F. Zenke, and P. Chambon. 1984. Stimulation of in vitro transcription from the SV40 early promoter by the enhancer involves a specific *trans*-acting factor. *EMBO J.* **3:** 3129.

The Use of Expression Selection for the Isolation of Cellular Enhancer Sequences

M. Fried, T. Williams, V. von Hoyningen-Huene, M. Ford, B. Davies, M. Griffiths, C. Thacker, and C. Passananti

Department of Tumour Virus Genetics
Imperial Cancer Research Fund, Lincoln's Inn Fields
London WC2A 3PX, England

A number of sequences required for efficient gene expression have been identified by their association with a cloned gene. We have used an alternative approach to clone expression sequences directly (Fried et al. 1983; Ford et al. 1985). This relies on their ability to reactivate a test gene devoid of its own expression sequences. In this report we summarize our results using this "expression selection" technique.

RESULTS AND DISCUSSION

The expression selection procedure for the isolation of cellular expression sequences is outlined in Figure 1. The test gene used is the polyoma virus (Py) early transforming region, which contains the complete coding sequences for the Py oncogene middle T antigen (Tooze 1981). However, this gene is greatly inhibited in its transforming activity since it lacks the Py enhancer element (de Villiers and Schaffner 1981) required for efficient early region gene expression. Restricted cellular DNA is ligated to the linearized Py vector, the ligated mixture is transfected onto rat cells, and transformed colonies are isolated. The transformed cells are assessed for reactivation of the Py genes by the production of Py T antigens and Py-specific mRNAs. The initiation of the RNA at the bona fide Py cap site suggests that reactivation is mediated by a cellular enhancer element. The Py sequences and adjacent cellular DNA are cloned from transformed cells containing a single viral insert by using a Py probe (secondary transfections may be required to obtain cells containing a single viral insert). Fragments of the cloned cellular DNA can be tested for their ability to stimulate the expression of a test gene in long- and short-term enhancer assays. Possible companion

68

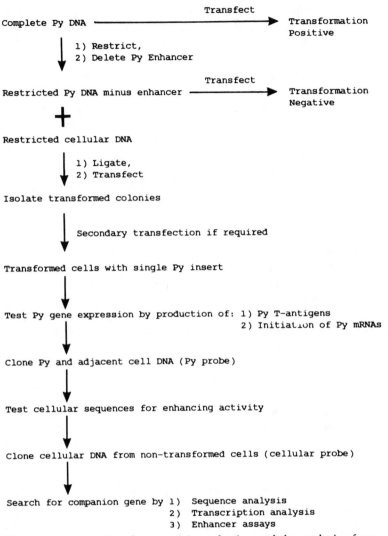

Figure 1 A procedure for expression selection and the analysis of companion genes. Modified from Fried et al. (1983) and Ford et al. (1985).

genes associated with the expression sequences can be cloned from nontransformed cellular DNA, using the previously cloned expression sequences as a probe, and further analyzed.

A 140-bp cellular sequence from mouse embryo DNA isolated by this procedure (Fried et al. 1983) was found to have enhancer properties, being capable of stimulating expression in an orientation-independent manner, when placed either 5' or 3' to a test gene. In the transfected transformed cell, the Py early region mRNAs were found to be initiated at the bona fide Py cap site (Fried et al. 1983), suggesting reactivation was via a cellular enhancer. This mouse expression sequence, MES-1 (formerly called H2), has been used as a probe to clone a 19-kb fragment from nontransformed mouse embryo DNA (T. Williams and M. Fried, pers. comm.) in order to analyze the sequence and any possible companion genes associated with MES-1. MES-1 contains a direct repeat of 10 bp but shares no homology with other previously identified enhancer consensus sequences (Fried et al. 1983). The sequence both upstream and downstream of MES-1 contains no obvious transcriptional signal elements (e.g., a TATA box). However, the sequence 5' to and including MES-1 is very GC rich. An 800-bp fragment from the 19-kb clone containing the MES-1 sequences at one end also shows orientation-independent enhancer activity. The region in the vicinity of MES-1 is highly transcriptionally active, producing at least three mRNAs. One mRNA has been found to be down-regulated in undifferentiated teratocarcinoma cells and another, which is quite abundant, appears to be a member of an unidentified gene family. Two of the mRNAs, which are transcribed in opposite directions, have been found to overlap by 120 bp at their 3' ends, suggesting some form of gene control. The association of MES-1 with one or more of these cellular mRNAs is being investigated further.

A second sequence, MES-2, was isolated from F9 teratocarcinoma DNA. MES-2, as well as the adjacent Py sequences, was found within a large inverted duplication that is amplified to between 20-fold and 40-fold in the cellular DNA from the transformed-cell colony arising after transfection (Ford et al. 1985). In these cells RNA was initiated at the bona fide Py cap site. The 4.7-kb fragment containing MES-2 has only weak enhancing activity when placed 5' to a test gene and transfected onto rat fibroblasts. The strongly transformed phenotype of the original transformed cell is probably a result of the 20-fold to 40-fold amplification of MES-2 (Ford et al. 1985). We are presently testing whether the enhancer activity of MES-2 is intrinsically weak or is reduced because of other reasons, e.g., being assayed in an inappropriate cell type.

After transfection using teratocarcinoma DNA, four other transformed cells lines that expressed the Py T antigens and appeared to contain multiple copies of Py DNA were isolated. When Hirt supernatants (Hirt 1967), which selectively isolate extrachromosomal DNA, were taken from all four cell lines, families of related circular molecules containing Py sequences were detected. In addition, these cells appeared to contain integrated Py sequences. In each line the circles were composed of a different basic unit, each of which was subsequently cloned. Preliminary analysis indicated that these circles are composed only of viral sequences. Cellular DNA sequences have not been detected, at least in the region where the original Py vector has been joined to form the circles. The Py vector fragment contained the core sequence for Py DNA replication but was incapable of replication since it lacked the Py enhancer sequences (Luthman et al. 1982; Muller et al. 1983). The circles may be replicated either as a result of the formation of a new enhancer element (viral or cellular sequences) or as a result of a *trans*-acting factor in a mutant cell type. Alternatively, they may be generated by excision from integrated Py sequences. These possibilities are currently under investigaton.

REFERENCES

de Villiers, J. and W. Schaffner. 1981. A small segment of polyoma virus DNA enhances the expression of a cloned β-globin gene over a distance of 1400 base pairs. *Nucleic Acids Res.* **9:** 6251.

Ford, M., B. Davies, M. Griffiths, J. Wilson, and M. Fried. 1985. Isolation of a gene enhancer within an amplified inverted duplication after "expression selection." *Proc. Natl. Acad. Sci.* **82:** 3370.

Fried, M., M. Griffiths, B. Davies, G. Bjursell, G. LaMantia, and L. Lania. 1983. Isolation of cellular DNA sequences that allow expression of adjacent genes. *Proc. Natl. Acad. Sci.* **80:** 2117.

Hirt, B. 1967. Selective extraction of polyoma DNA from infected mouse cell cultures. *J. Mol. Biol.* **26:** 365.

Luthman, H., M.-G. Nilsson, and G. Magnusson. 1982. Non-contiguous segments of the polyoma genome required in cis for DNA replication. *J. Mol. Biol.* **161:** 533.

Muller, W.J., C. Mueller, A.-M. Mes, and J.A. Hassell. 1983. Polyomavirus origin for DNA replication comprises multiple genetic elements. *J. Virol.* **47:** 586.

Tooze, J., ed. 1981. *Molecular biology of tumor viruses.* 2nd edition, revised: *DNA tumor viruses.* Cold Spring Harbor Laboratory, Cold Spring Harbor, New York.

Transcriptional Control Sequence–binding Factors Require Interaction for T-antigen-mediated SV40 Late Gene Expression

J. Brady, M. Loeken, and G. Khoury

National Cancer Institute, National Institutes of Health
Bethesda, Maryland 20205

DNA viruses have proven to be valuable models for the elucidation of eukaryotic transcriptional control elements. The SV40 72-bp repeat, an enhancer element capable of increasing the level of transcription 100-fold to 1000-fold, provided an introduction for the subsequent discovery of cellular enhancer elements (see Gluzman and Shenk 1983; Khoury and Gruss 1983). Among the properties of these regulatory elements is their ability to function in a position- and orientation-independent fashion as well as to induce the expression of heterologous genes. The organization of the SV40 transcriptional control region has been well established by nucleotide sequencing as well as through biological, genetic, and biochemical studies. The location of the SV40 tandem repeats between the early and late SV40 genes suggests that it might function equally well for both transcriptional units. SV40 late genes, however, are normally expressed only after the synthesis of sufficient amounts of the early gene product, T antigen, and the initiation of viral DNA replication. Furthermore, late gene expression occurs efficiently only in permissive monkey kidney cells. These transcriptional properties of the late promoter present a conceptual problem in light of the bidirectional properties described for the prototype SV40 enhancer element.

Recently, we (Brady et al. 1984; Brady and Khoury 1985) and others (Hartzell et al. 1984; Keller and Alwine 1984) have found that SV40 late gene expression is activated by the SV40 early gene product, T antigen. RNA analysis, either by S1 nuclease or Northern blot techniques, demonstrates that activation occurs at the transcriptional level (Keller and Alwine 1984; Brady and Khoury 1985). Studies in our laboratory using deletion and point mutants have defined two important domains for the SV40 T-antigen-induced late gene expression. One of these includes T-antigen-binding sites I and II, while the other is located in the SV40 72-bp repeat. From these experiments, it was not possible to distinguish whether the transcriptional control sequences represented promoter elements and/ or binding sites for *trans*-acting factors. To address this point and to determine how the two upstream control elements interact to effect T-antigen-dependent *trans* activation, we have used template competition analysis. In the presence of increasing levels of competitor DNA fragments, which are capable of binding limiting *trans*-acting factors, a decrease in expression from a fixed amount of template was observed. The methodology thus allows the definition of DNA domains that bind regulatory proteins present at limiting levels in the eukaryotic cell.

RESULTS

COS-1 cells, which constitutively express SV40 T antigen from a single integrated copy of the viral DNA, were transfected with increasing amounts of wild-type SV40 DNA. Cytosine arabinoside (25 μg/ml) was added to the culture media to prevent DNA replication. It was determined that 0.1 μg of SV40 template DNA provided 60% of the maximal VP-1 response in transfected 10-cm plates. No expression of the SV40 late genes was observed in control CV-1 monkey kidney cells. The fact that saturation can be obtained suggests that a limiting transcriptional factor(s) is present in the transfected COS-1 cells. In our initial template competition studies, the entire control region of SV40 containing the three T-antigen-binding sites and the enhancer region (Fig. 1A) was introduced at 1:1, 10:1, and 100:1 ratios in relation to the SV40 template DNA. The presence of the limiting factor was demonstrated by the quantitative reduction in late gene expression as measured by immunoblot analysis of the major late protein, VP-1 (Fig. 1B). The level of late gene expression was proportional to the ratio of template : competitor DNA (Fig. 1B,C).

To locate more precisely DNA sequences that are involved in binding of the *trans*-acting factor(s), in vivo competition studies were performed with a series of cloned DNA promoter fragments. A competition fragment that contains half of T-antigen-binding site II, T-antigen-binding site III (including the 21-bp repeats), and the 72-bp enhancer element was an ineffective competitor (Fig. 2B). Similar inability to compete was observed with fragments that contain the 72-bp repeats and portions of the 21-bp repeats (Fig. 2C,D). A competitor DNA that includes T-antigen-binding sites I, II, and III

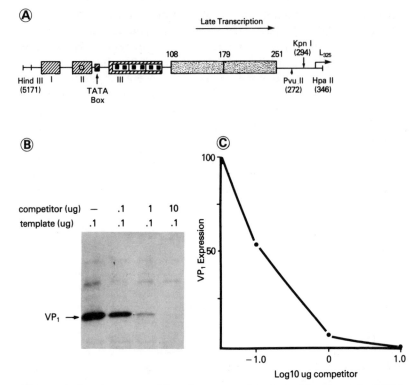

Figure 1 In vivo competition for *trans*-acting factor(s) present in COS-1 cells. COS-1 cells on a 10-cm plate were transfected with SV40 template DNA and competition plasmids by the calcium phosphate precipitation method. The transfection mixtures contained 0.1 μg of supercoiled SV40 template DNA plus a competitor DNA molecule containing the SV40 T-antigen-binding sites and late promoter (SV40 nt 5171–272) cloned into pBR322 (*A*). The competition plasmid lacked the complete coding region for the major late gene product VP-1. The competitor plasmid DNA was added at a ratio of 1:1 (0.1 μg), 10:1 (1.0 μg), or 100:1 (10 μg) relative to template DNA (*B*). Transfected cultures were maintained in Dulbecco's minimal essential medium with fetal calf serum (10%) and cytosine arabinoside (25 μg/ml). At 40 hr posttransfection, whole cell protein extracts were prepared and analyzed by immunoblot analysis using anti-SV40 VP-1 antisera (Brady et al. 1984). After autoradiography, bands corresponding to SV40 VP-1 were excised from the blot and ^{125}iodine cpm was determined (*C*). The level of SV40 late expression obtained by transfection of 0.1 μg SV40 DNA was set at 100%. (*B* and *C* reprinted, with permission, from Brady and Khoury 1985.)

(21-bp repeats) but lacks the SV40 72-bp repeats was able to function as a competitor (binding *trans*-acting factors), but the efficiency was reduced substantially (Fig. 2E). We conclude that for efficient

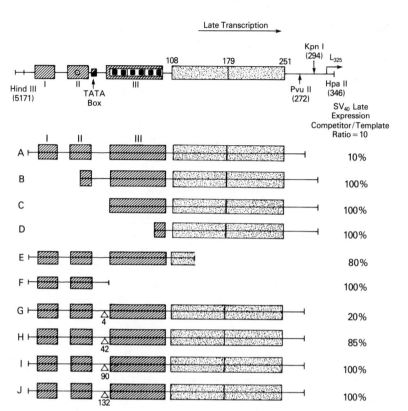

Figure 2 SV40 late gene expression in COS-1 cells: Delimitation of DNA-binding site(s) for *trans*-acting factor(s). In vivo competition assays were set up as described in the legend to Fig. 1. Competition plasmids were added at a 10:1 (1.0 μg) ratio to SV40 template DNA (0.1 μg). Whole cell protein extracts were subjected to SDS-polyacrylamide gel electrophoresis, transferred to nitrocellulose, and analyzed by immunoblot analysis using anti-SV40 VP-1 antisera. The SV40 sequences contained in the competition plasmids are as follows: (A) 5171–272, (B) 1–294, (C) 41–294, (D) 95–272, (E) 5171–128, (F) 5171–37, (G) 5171–272 (4-bp insertion at *Nco*I site), (H) 5171–272 (42-bp insertion at *Nco*I site), (I) 5171–272 (90-bp insertion at *Nco*I site), (J) 5171–272 (132-bp insertion at *Nco*I site). The level of SV40 late expression was determined by quantitation of bound [125]iodine cpm. The level of expression obtained with 0.1 μg SV40 template DNA, in the absence of competitor DNA, represents 100%.

competition, a fragment must contain both of the transcriptional control elements for T-antigen-activated late transcription (i.e., T-antigen-binding sites I and II and the 72-bp repeats). In a separate experiment, it was shown that when the two domains were introduced as potential competitors on separate plasmids (Fig. 2C,F), no competition was observed. Thus, the two regulatory domains must maintain a *cis* relationship to one another.

A set of competitor fragments that consists of the late SV40 control region containing inserts at the *Nco*I site (SV40 nt 37) were also used in the in vivo competition assay (Fig. 2G,H,I,J). The results of these competition experiments demonstrate that increasing the distance between the two transcriptional domains reduces the ability of the DNA fragment to bind limiting *trans*-acting factors required by the SV40 template. Insertion of four nucleotides had a rather minimal effect on the competitor fragment, reducing its efficiency by about 20% (Fig. 2G). Insertion of 42 nucleotides effectively eliminated the ability of the competitor fragment to reduce VP-1 expression from the template (Fig. 2H). Similar results were obtained with insertion mutants containing larger segments of DNA separating the two late transcriptional control domains (Fig. 2I,J). We conclude that not only are these two domains required in *cis*, but also that the spatial relationship between them is critical for binding limiting late transcriptional factor(s).

DISCUSSION

Using in vivo competition assays, we have extended our analysis of the regulation of SV40 late gene expression. Our data suggest that both T-antigen-binding sites I and II and the SV40 72-bp repeats bind transcriptional factors required for T-antigen *trans* activation. The fact that these two binding domains are ineffective alone, or when spatially separated from each other, suggests that the proteins associating with these domains may physically interact to enhance transcription from the SV40 late promoter. This is the first experimental evidence of which we are aware that suggests the potential importance of protein-protein interaction in polymerase II transcription. Alternatively, the SV40 72-bp repeats may be required to maintain an open chromatin structure to allow efficient binding of *trans*-acting factors.

Our data demonstrate that the SV40 72-bp repeat is important for T-antigen-mediated late gene expression. The mechanism by which the 72-bp enhancer element functions, however, may be different for early and late gene expression. The SV40 early enhancer can be activated by transcriptional factors that specifically recognize the 72-bp repeats (Schöler and Gruss 1984) and are endogenous to many cell types. The function of the 72-bp transcriptional control sequence that contributes to late transcription is inducible by T antigen. In addition, binding of the late transcription factor to the enhancer element apparently requires T antigen or another transcriptional factor that associates with T-antigen sites I and II. It has not been determined whether the SV40 72-bp repeats function as a classical enhancer (position- and orientation-independent) for late gene expression.

The inducibility of the SV40 enhancer by T antigen for its role in late transcription may be similar to the induction of cellular transcriptional regulatory sequences during differentiation or to the activation of certain genes in response to hormone-receptor complexes (e.g., see Chandler et al. 1983). This concept of "inducible enhancers" may apply to genes that require factors for their activation that are produced only at specific times in a cell cycle. We anticipate that analysis of the mechanism of induction of SV40 late gene expression by T antigen and other transcriptional factors will provide insight into the regulation of these genes.

ACKNOWLEDGMENTS

We express our appreciation to J. Duvall for expert technical assistance and to M. Priest for preparation of the manuscript.

REFERENCES

Brady, J. and G. Khoury. 1985. *Trans*-activation of the SV40 late transcription unit by T-antigen. *Mol. Cell. Biol.* **5**: 1391.

Brady, J., J. Bolen, M. Radonovich, N.P. Salzman, and G. Khoury. 1984. Stimulation of simian virus 40 late expression by simian virus 40 tumor antigen. *Proc. Natl. Acad. Sci.* **81**: 2040.

Chandler, V.L., B.A. Maler, and K.R. Yamamoto. 1983. DNA sequences bound specifically by glucocorticoid receptor *in vitro* render a heterologous promoter hormone responsive *in vivo*. *Cell* **33**: 489.

Gluzman, Y. and T. Shenk, eds. 1983. *Current communications in molecular biology: Enhancers and eukaryotic gene expression*. Cold Spring Harbor Laboratory, Cold Spring Harbor, New York.

Hartzell, S.W., B.J. Byrne, and K.N. Subramanian. 1984. The simian virus 40 minimal origin and the 72-base-pair repeat are required simultaneously for efficient induction of late gene expression with large tumor antigen. *Proc. Natl. Acad. Sci.* **81**: 6335.

Keller, J.M. and J.C. Alwine. 1984. Activation of the SV40 late promoter: Direct effects in the absence of viral DNA replication. *Cell* **36**: 381.

Khoury, G. and P. Gruss. 1983. Enhancer elements. *Cell* **33**: 313.

Schöler, H.R. and P. Gruss. 1984. Specific interaction between enhancer-containing molecules and cellular components. *Cell* **36**:403.

Interaction of Cellular Factors with the SV40 and Polyoma Enhancers

S. Cereghini, J. Piette, M.-H. Kryszke, and M. Yaniv

Department of Molecular Biology, Pasteur Institute
75015 Paris, France

SV40 and polyoma viruses contain transcription enhancer sequences upstream of their early and late promoters. These sequences were shown to activate transcription from cellular or viral promoters when placed either 5' or 3' to a transcription unit in both possible orientations (Gruss 1984). Even though these two viruses show extensive homology in the coding regions and in the structure of the palindromic origin of replication, they differ markedly in the position and sequence of their enhancer elements. In SV40 the 72-bp repeat element that constitutes the major part of the enhancer starts at position −112 relative to the early cap site with the origin of replication roughly coinciding with the cap site. In polyoma the enhancer element starts roughly at position −180 relative to the cap site of early mRNA with the origin sequence inserted between the early promoter and its enhancer elements. Furthermore, the SV40 enhancer contains a duplication of 72 bp with active sequences extending probably beyond up to the *Pvu*II site (position 273), whereas wild-type polyoma of the A2 or A3 strains contains only unique sequences composed of two distinct enhancer elements with different cell specificity located between the *Bcl*I and *Pvu*II (enhancer A) sites and *Pvu*II-*Pvu*II sites (enhancer B), respectively (Herbomel et al. 1984 and Fig. 1). In some wild-type polyoma strains like the Toronto large plaque, the core of enhancer element A is duplicated (Ruley and Fried 1983), whereas in others like 2PT-A, the core of element B is duplicated (G. Mandel et al., pers. comm.). In both viruses the enhancer sequences are associated with a nucleosome-free, nuclease-hypersensitive DNA sequence (Saragosti et al. 1980; Herbomel et al. 1981). We suggested that the alternation of resistant and hypersensitive sites along the region from the origin to roughly the *Kpn*I site of SV40 can be the result of the binding of certain cellular proteins to these sequences. In fact, the same pattern was

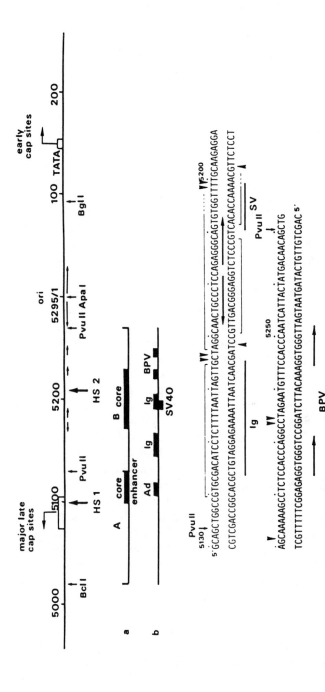

Figure 1 Control sequences of polyoma virus transcription and replication (*top*) and DNase I protection pattern after formation of DNA-protein complexes in vitro (*bottom*). HS1 and HS2 are DNase I hypersensitive sites observed in vivo. The core sequences of the polyoma enhancers A and B are indicated. The top strand corresponds to the early coding sequences. Arrows along the sequence indicate inverted and direct repeats. Homologies with IgG (Ig), SV40 (SV), and bovine papilloma virus (BPV) enhancer sequences are indicated. Protected sequences are given in brackets, and hypersensitive sites are marked by arrowheads.

79

seen when plasmid DNA containing only the noncoding region of SV40 was replicated in COS cells, or after shift to 40°C of cells infected with *tsA* mutants of SV40 (Cereghini and Yaniv 1984). In this communication we demonstrate the mapping by in situ footprinting of protein-DNA contacts in the SV40 enhancer and the isolation of mouse proteins that bind to the polyoma B enhancer in vitro.

SV40 Enhancer

Nuclei from SV40-infected cells were digested for various lengths of time with micrococcal nuclease, followed by extraction of total DNA. Alternatively, SV40 minichromosomes were extracted from infected nuclei and treated with the nuclease. DNA samples were cleaved with either *Hind*III or *Bgl*I, and the DNA products were fractionated on a denaturing sequencing gel. Samples of pure SV40 DNA cleaved with nuclease and chemical degradation products of linear SV40 (cleaved with either *Hind*III or *Bgl*I) were run in parallel. After transfer to nylon membranes, the viral DNA was revealed by single-strand DNA probes abutting the restriction site (Church and Gilbert 1984). Two identical segments of protected DNA were observed with probes hybridizing with the late strand along each of the 72-bp repeats (summarized in Fig. 2). The first segment around the *Eco*RII site (nt 153–169 in the first repeat and 225–241 in the second repeat) was totally protected, with a hypersensitive site in the middle of the protected sequences at positions 161 and 223. The second segment around the *Sph*I site (positions 121–139 and 193–212 in the first and second repeats, respectively) was only partially protected, again with enhanced cleavage in the middle of the protected sequence. This protection pattern is much more pronounced along the late coding strand than it is along the early coding strand. We define the *Sph*I binding protein as enhancer-binding protein 1 or EBP1, and the *Eco*RII binding protein we call EBP2. In addition, about 15 bp (nt 261–275 on the late strand and 265–277 in the early strand) were fully protected around the *Pvu*II site, a region that is important for the enhancer function, although it is outside of the 72-bp repeat. We define it as EBP3 binding site. The three binding sites contain repeated sequences (of 5, 7, and 4, respectively) situated 10 bp apart. These repeats could be the site of interaction of two monomeric subunits contacting the same side of the double helix. The association of these proteins with the viral DNA sequences is relatively unstable. Extraction of minichromosomes in isotonic conditions and their purification by sucrose gradient centrifugation in 0.12 M salt results in an almost total loss of the pro-

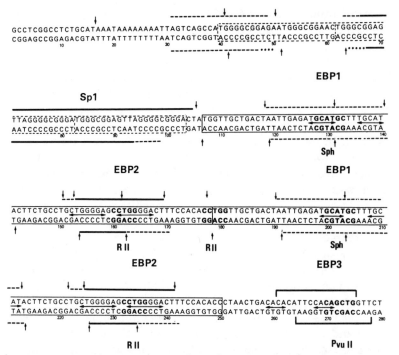

Figure 2 DNA-protein contacts along the promoter-enhancer sequences of SV40. The nucleotide sequence of SV40 (residues 1–280 in the BBB system). The upper strand is the late coding strand. The two repetitions of 21 bp and 72 bp as well the 22-bp element are boxed. The *Sph*I, *Eco*RII, and *Pvu*II restriction sites are in bold type. Full protection observed in nuclei after micrococcal nuclease digestion is indicated by a continuous line, and partial protection is indicated by a dashed line above (late strand) or below (early strand) the sequence. (→) Hypersensitive sites. The protection along the 21-bp repeats is tentatively attributed to the Sp1 transcription factor of Dynan and Tjian (1983). Repeated motifs in the contact sites of enhancer-binding proteins are indicated.

tected pattern. In addition to protection, several sites show increased sensitivity; these sites also disappear after the purification of the minichromosomes.

In vitro mutagenesis experiments have shown that both the *Sph*I site (A. Rich et al., pers. comm.) and the residues around the *Eco*RII site (including also the GTGGAAAG consensus sequence) are crucial for the function of the viral enhancer (Weiher et al. 1982; Herr and Gluzman 1985). In addition to this protection, we see protection of the two 21-bp repeats, probably by the Sp1 transcription factor (Dynan and Tjian 1983; Gidoni et al. 1984). Here also the maximal protection is observed along the late coding strand, and the protein is lost relatively easily.

Polyoma Enhancer

In this case we took a slightly different approach. We searched for mouse proteins present in noninfected cells that bind to the enhancer sequences of polyoma. To detect DNA-protein interaction, we employed the gel retardation technique of Garner and Revzin (1981) as modified by Strauss and Varshavsky (1984). Nuclei from mouse fibroblasts were extracted at different salt concentrations, and the concentrated fractions were incubated with end-labeled radioactive DNA fragments in the presence of a large excess of nonspecific DNA (plasmid or salmon sperm). We found specific retardation of DNA fragments containing the B enhancer of polyoma. Addition of excess nonradioactive DNA containing the B enhancer competed efficiently with the radioactive fragment, whereas fragments containing the A enhancer of polyoma or the enhancer-promoter origin of SV40 did not compete. Several intermediate bands migrating faster than the complete complex were observed in the gel. A plot of their mobility versus the hypothetical molecular weight of the complex (Fried and Crothers 1981) suggests that the enhancer-protein complex contains at least two different proteins, a large one and three identical subunits of a smaller polypeptide. By DNase I footprinting, we compared the cleavage sites in the complexed and noncomplexed radioactive DNA fragment. As shown in Figure 2, protein-DNA contacts were detected along sequences that were defined previously as the core of element B (Herbomel et al. 1984). It included the Weiher and Gruss consensus sequence, the GC-rich palindrome, and a sequence homologous to the IgG heavy-chain enhancer, which was protected only on one strand. Several hypersensitive sites were also observed.

CONCLUSIONS

We have shown both for SV40 and polyoma that several proteins were associated with the active enhancer sequences. The binding sites detected in the SV40 enhancer sequences coincide with nucleotides that were shown to be crucial for enhancer function. Deletions and point mutants studied by Chambon's group have outlined the importance of the 5' and 3' domains of the 72-bp repeat and of residues located close to the *Pvu*II site for enhancer function (Moreau et al. 1981; Sassone-Corsi et al. 1985; P. Chambon, pers. comm.). The work of Weiher et al. (1982) and Herr and Gluzman (1985) have outlined the importance of G + C residues around the

*Eco*RII site. Furthermore, alternating purine-pyrimidine residues around the *Sph*I and *Pvu*II sites are important for enhancer activity. Similarly, for polyoma the region protected against DNase I in the protein-DNA complex contains sequences common to polyoma, SV40, and IgG enhancers (Banerji et al. 1983; Gillies et al. 1983) and to the GC-rich palindrome that is conserved in all the rearranged host-range mutants of polyoma (Vasseur et al. 1982).

The proteins that bind to the viral enhancers are relatively abundant in the cells. SV40-infected cells can contain up to 100,000 copies of viral DNA in their nuclei in the beginning of the late phase. The entire population seems to contain bound EBP2. Similarly, in the initial nuclear extracts of 3T6 cells, we can already detect the specific binding to enhancer B, showing that these proteins are quite abundant. What is astonishing is that the complexes formed between these proteins and the DNA are quite salt sensitive, and in the case of SV40 the mere purification of the minichromosomes on a sucrose gradient resulted in their almost total dissociation.

Only a small fraction of the intranuclear viral minichromosomes are effectively transcribed at any moment. We propose that the enhancer-binding proteins create the framework for an efficient initiation of transcription. The rate of initiating RNA polymerase is dependent on the availability of the enzyme and of other transcription factors that bind to the promoter sequences (e.g., TATA-binding protein). This rate is not limited by the availability of Sp1 since most of the viral minichromosomes contain a protein bound to the 21-bp repeat, which is presumably Sp1. Perhaps the binding of the transcription initiation factor will create a salt-resistant initiation complex similar to the complexes discussed by Brown (1984).

Our data on SV40 show that EBP1 and Sp1 occupy neighboring sequences. It is not unlikely that they will form protein-protein contacts. One can wonder whether some components of the enhancer function are active only in close proximity to the promoter while others are also seen at a distance. This may explain the discrepancy between different groups of the distance effect on enhancer strength. The sequence protected by the EBP2 of SV40 includes a TGG sequence identified also in the four homologous clusters in the IgG enhancer that are probably involved in its function (Ephrussi et al. 1985).

REFERENCES

Banerji, J., L. Olson, and W. Schaffner. 1983. A lymphocyte-specific cellular enhancer is located downstream of the joining region in immunoglobulin heavy chain genes. *Cell* **33**: 729.

Brown, D.D. 1984. The role of stable complexes that repress and activate eucaryotic genes. *Cell* 37: 359.

Cereghini, S. and M. Yaniv. 1984. Assembly of transfected DNA into chromatin: Structural changes in the origin-promoter-enhancer region upon replication. *EMBO J.* 3: 1243.

Church, G.M. and W. Gilbert. 1984. Genomic sequencing. *Proc. Natl. Acad. Sci.* 81: 1991.

Dynan, W.S. and R. Tjian. 1983. The promoter specific transcription factor Sp1 binds to upstream sequences in the SV40 early promoter. *Cell* 35: 79.

Ephrussi, A., G.M. Church, S. Tonegawa, and W. Gilbert. 1985. B lineage–specific interactions of an immunoglobulin enhancer with cellular factors in vivo. *Science* 227: 134.

Fried, M. and D.M. Crothers. 1981. Equilibria and kinetics of *lac* repressor-operator interactions by polyacrylamide gel electrophoresis. *Nucleic Acids Res.* 9: 6505.

Garner, M.M. and A. Revzin. 1981. A gel electrophoresis method for quantifying the binding of proteins to specific DNA regions: Application to the components of the *Escherichia coli* lactose operon regulatory system. *Nucleic Acids Res.* 9: 3047.

Gidoni, D., W.S. Dynan, and R. Tjian. 1984. Multiple specific contacts between a mammalian transcription factor and its cognate promoters. *Nature* 312: 409.

Gillies, S.D., S.L. Morrison, V.T. Oi, and S. Tonegawa. 1983. A tissue specific transcription enhancer element is located in the major intron of a rearranged immunoglobulin heavy chain gene. *Cell* 33: 717.

Gruss, P. 1984. Magic enhancers. *DNA* 3: 1.

Herbomel, P., B. Bourachot, and M. Yaniv. 1984. Two distinct enhancers with different cell specificities coexist in the regulatory region of polyoma. *Cell* 39: 653.

Herbomel, P., S. Saragosti, D. Blangy, and M. Yaniv. 1981. Fine structure of the origin proximal DNase I hypersensitive region in wild type and EC mutant polyoma. *Cell* 25: 651.

Herr, W. and Y. Gluzman. 1985. Duplications of a mutated SV40 enhancer restore its activity. *Nature* 313: 711.

Moreau, P., R. Hen, B. Wasylyk, R. Everett, M.P. Gaub, and P. Chambon. 1981. The SV40 72 base pair repeat has a striking effect on gene expression both in SV40 and other chimeric recombinants. *Nucleic Acids Res.* 9: 6047.

Ruley, H. and M. Fried. 1983. Sequence repeats in a polyoma virus DNA region important for gene expression. *J. Virol.* 47: 233.

Saragosti, S., G. Moyne, and M. Yaniv. 1980. Absence of nucleosomes in a fraction of SV40 chromatin between the origin of replication and the region coding for the late leader RNA. *Cell* 20: 65.

Sassone-Corsi, P., A. Wildeman, and P. Chambon. 1985. A *trans*-acting factor is responsible for the SV40 enhancer activity *in vitro*. *Nature* 313: 458.

Strauss, F. and A. Varshavsky. 1984. A protein binds to a satellite DNA repeat at three specific sites that would be brought into mutual proximity by DNA folding in the nucleosome. *Cell* 37: 889.

Vasseur, M., M. Katinka, P. Herbomel, M. Yaniv, and D. Blangy. 1982. Physical and biological features of polyoma virus mutants able to infect embryonal carcinoma cell lines. *J. Virol.* **43:** 800.

Weiher, H., M. Konig, and P. Gruss. 1982. Multiple point mutations affecting the simian virus 40 enhancer. *Science* **219:** 626.

Tissue-specific Footprints of the Immunoglobulin Heavy-chain Enhancer In Vivo

A. Ephrussi,* G.M. Church, † W. Gilbert, † and S. Tonegawa*

*Center for Cancer Research and Department of Biology
Massachusetts Institute of Technology
Cambridge, Massachusetts 02139

†Biogen Research Corporation, Cambridge, Massachusetts 02142

Immunoglobulin heavy-chain genes contain a tissue-specific enhancer element (Banerji et al. 1983; Gillies et al. 1983; Neuberger 1983). This transcriptional enhancer element is located in the intron separating the J-region cluster and the first $C\mu$ exon, upstream of the switch region. Thus, regardless of which V, D, J, and C gene segments are selected for use by rearrangement (Tonegawa 1983), in a given cell the enhancer element is always present within the assembled heavy-chain gene.

The immunoglobulin heavy-chain enhancer is tissue specific; i.e., in stable and transient expression transfection assays, it stimulates transcription of the immunoglobulin gene and of heterologous genes and promoters only in cells of the B lineage. This element is effective when positioned in either orientation several kilobases upstream or downstream of the promoter on which it acts.

In an attempt to determine whether cellular factors interact with the enhancer in a tissue-specific fashion and to define precisely the sequences that might be involved within the enhancer region, we carried out dimethylsulfate (DMS) protection experiments (Gilbert et al. 1976) on intact cells (Ephrussi et al. 1985) and on nuclei (Church et al. 1985). Although DMS methylates all guanine (G) and adenine (A) residues of naked DNA, proteins that interact with specific DNA sequences can prevent or increase DMS methylation of G residues at the site of protein binding. We treated cells or nuclei with DMS, allowing a partial methylation of G residues to occur. The DNA was then extracted and the region containing the immunoglobulin enhancer was visualized by genomic sequencing (Church and Gilbert 1984).

We examined the G residues of the immunoglobulin heavy-chain enhancer and sequences flanking it, in cells of the B lineage (myelomas J558L, MOPC104E, and MPC-11; a B lymphoma, A20-2J; and an Abelson virus–transformed pre-B cell, RAW8), and compared their patterns of G reactivity to DMS with those of cells not of the B lineage (Friend virus erythroleukemia cells, MEL; L cells; and two T lymphomas, S49 and BW5147).

Overall, from experiments on intact cells, the following picture emerges: The G residues of myeloma, B, and pre-B cells are virtually indistinguishable from one another and reveal a common pattern distinct from the pattern of non-B cells and of naked DNA. Over the nearly 1000 bases on the top and the 430 bases on the bottom strand that were examined, we found 28 protected and 4 enhanced G residues in B-lineage cells relative to non-B-lineage cells and naked DNA (Fig. 1). Nearly all of these tissue-specific alterations in reactivity are localized precisely to the small region defined as being essential for transcriptional enhancer function (Banerji et al. 1983; Gillies et al. 1983).

Most of these alterations lie within four clusters that contain sequences homologous to the octamer $\frac{\text{CAGGTGGC}}{\text{GTCCACCG}}$ (Fig. 2). The first cluster matches this consensus octamer at six out of eight positions; the other three match this octamer at seven out of eight positions. When present within these sequences, in cells of the B lineage, G residues at positions 3, 4, 6, and 8 are always protected; G residues at position 1 can be either protected or normal, and at position 7, either protected, enhanced, or normal.

To investigate the stabilities of these tissue-specific effects, we carried out experiments varying the monovalent cation concentration of the incubation buffer prior to treatment of B- and non-B-lineage cell nuclei with DMS. We examined sites 3 and 4 in these two categories of cells and in naked DNA. The pattern of tissue-specific protections and enhancements is found at salt concentrations of 10, 75, and 150 mM but is not observed at 250 and 500 mM salt, where the pattern of reactivity of the G residues of B-lineage cells becomes identical to that of non-B cells and naked DNA.

The simplest interpretation of these results is that proteins interact with the immunoglobulin heavy-chain enhancer at the octamers CAGGTGGC and that these interactions are tissue specific. These interactions are correlated with enhancer function: The cells in which the protections and enhanced reactivities occur are all cells of the B lineage that express immunoglobulin mRNA.

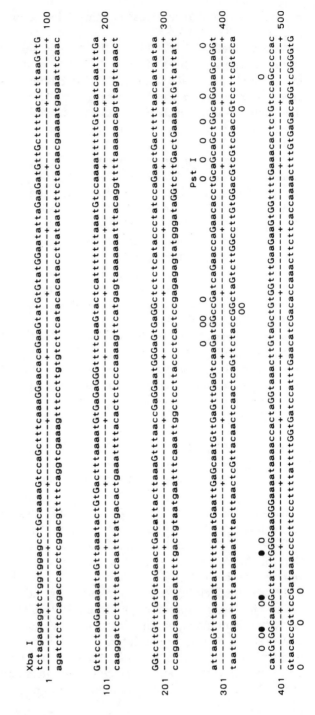

88

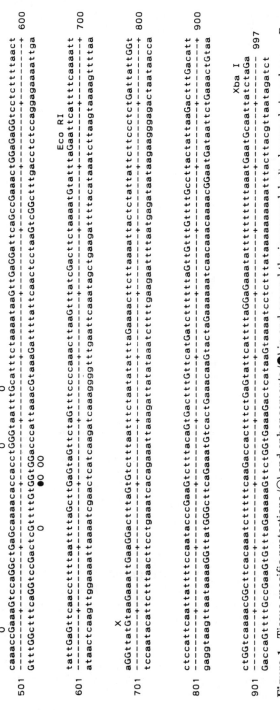

Figure 1 Tissue-specific protections (○) and enhancements (●) in and around the immunoglobulin heavy-chain enhancer. Residues whose reactivities to DMS were examined are capitalized. (Reprinted, with permission, from Ephrussi et al. 1985.)

89

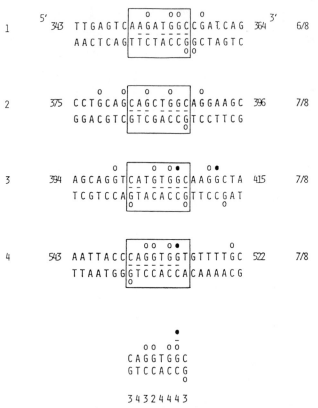

Figure 2 Sequence comparison of the four clusters of the B-lineage-specific protections (○) and enhancements (●). An octamer consensus sequence is indicated at the bottom. Within the enhancer, clusters 1, 2, and 3 occur in one orientation and cluster 4 occurs in the opposite orientation. (Reprinted, with permission, from Ephrussi et al. 1985.)

ACKNOWLEDGMENTS

This work was supported by National Institutes of Health grants AI-17879-03 (to S.T.) and GM-09541-22 (to W.G.), Biogen NV (W.G.), NIH training grant 2T-32CA09255 (to A.E.), and American Cancer Society grant ACS-NP-327A (to S.T.).

REFERENCES

Banerji, J., L. Olson, and W. Schaffner. 1983. A lymphocyte-specific cellular enhancer is located downstream of the joining region in immunoglobulin heavy chain genes. *Cell* **33:** 489.

90

Church, G.M. and W. Gilbert. 1984. Genomic sequencing. *Proc. Natl. Acad. Sci.* **81:** 1991.

Church, G.M., A. Ephrussi, W. Gilbert, and S. Tonegawa. 1985. Cell-type-specific contacts to immunoglobulin enhancers in nuclei. *Nature* **313:** 798.

Ephrussi, A., G.M. Church, S. Tonegawa, and W. Gilbert. 1985. B lineage–specific interactions of an immunoglobulin enhancer with cellular factors *in vivo. Science* **227:** 134.

Gilbert, W., A. Maxam, and A.D. Mirzabekov. 1976. DNA contacts between the *lac* repressor and DNA revealed by methylation. *Proc. Alfred Benzon Symp.* **9:** 139.

Gillies, S.D., S.L. Morrison, V.T. Oi, and S. Tonegawa. 1983. A tissue-specific transcription enhancer element is located in the major intron of a rearranged immunoglobulin heavy chain gene. *Cell* **33:** 717.

Neuberger, M.S. 1983. Expression and regulation of immunoglobulin heavy chain gene transfected into lymphoid cells. *EMBO J.* **2:** 1373.

Tonegawa, S. 1983. Somatic generation of antibody diversity. *Nature* **302:** 575.

Sequences Governing Immunoglobulin Transcription In Vitro

R. Sen and D. Baltimore
Whitehead Institute for Biomedical Research
Cambridge, Massachusetts 02142

Experiments employing DNA transfection into cells in tissue culture in recent years helped to define DNA sequences that are critical for the expression of genes. In the case of immunoglobulin genes, at least two regulatory sequences have been defined. Both the heavy-chain genes as well as the *x* light-chain genes contain a regulatory element in the large intron between the J and C regions that has all the properties of classical viral enhancers (Banerji et al. 1983; Gillies et al. 1983; Picard and Schaffner 1983; Queen and Baltimore 1983). In addition to activating genes at large distances and in either orientation, the immunoglobulin enhancers are tissue specific (i.e., they function only in cells of B lymphoid lineage). A second regulatory element in the classical promoter region has been postulated (Parslow et al. 1984) in all immunoglobulin genes and verified for the *x* genes (Bergman et al. 1984; Falkner and Zachau et al. 1984).

To understand the molecular mechanisms by which these sequences regulate immunoglobulin expression, we have initiated a project to study the transcription of these genes in vitro. Soluble cell-free extracts that accurately initiate transcription by RNA polymerase II have been developed (Weil et al. 1979; Manley et al. 1980), and in these both upstream effects (e.g., Tsuda and Suzuki 1981; Grosschedl and Birnstiel 1982; Hen et al. 1982; Hansen and Sharp 1983; Jove and Manley 1984) and enhancer effects (Sassone-Corsi et al. 1984; Sergeant et al. 1984; Wildeman et al. 1984) have been observed. Furthermore, these extracts have also yielded specific cellular factors that interact with the TATA box region (Davison et al. 1983; Parker and Topol 1984a) and upstream sequence elements (Dynan and Tjian 1983; Parker and Topol 1984b).

RESULTS
The plasmid p*x* contains a truncated MOPC41 *x* gene retaining 110 bp upstream from the cap site and approximately 2 kb of the ge-

nomic sequence. $p\varkappa/E\varkappa$ and $p\varkappa/E\mu$ were generated from $p\varkappa$ by inserting either an 800-bp fragment from the $J_\varkappa/C_\varkappa$ intron containing the $\varkappa$ enhancer or a 700-bp fragment from the $J_\varkappa/C_\varkappa$ intron carrying the heavy-chain enhancer into the unique HindIII site. The plasmids were linearized by cleaving with Sal in order to generate a 2.3-kb runoff transcript.

Our initial experiments utilized whole cell extracts (Manley et al. 1980) made from the mouse myeloma MPC-11. We have been able to fractionate this extract and reconstitute a transcription system only from B cells; however, differential transcription from templates containing or not containing the $\varkappa$ enhancer was not observed. Since a serious drawback of a fractionated and reconstituted transcription system is the possibility that important regulatory factors may be absent from the final fractions (e.g., Hansen and Sharp 1983), we attempted to derive whole cell extracts from B lymphoid cells that would be transcriptionally active. Two such extracts have been derived to date from human B-lymphoma cell lines, RAMOS and EW-36 (provided by Dr. Gilbert Lenoir). Runoff analysis of transcription of the $\varkappa$ templates in these extracts as well as in a HeLa whole cell extract shows no dependence on the presence or absence of the enhancer element after a 60-min reaction. Various conditions were unsuccessfully tried to observe differential transcription of the two templates. The presence of polyethylene glycol (PEG), however, makes a significant difference in the transcriptional analysis. The experiments to be described were carried out by preincubating the DNA template with the whole cell extract in the presence of 5–6% PEG for 30–60 min, followed by initiating the transcription reaction with nucleotides and radioactive uridine triphosphate for 10 min. A chase with great excess of unlabeled nucleotides allows one to measure the extent of initiation during the pulse. Under these conditions we have reproducibly observed 4-fold to 10-fold decreased transcription of a template containing an enhancer as compared with one not containing the enhancer. The effect can be observed with linear templates as well as closed circular templates (Fig. 1). Simple mixing experiments where templates $p\varkappa$ and $p\varkappa/E\varkappa$ are transcribed individually as well as together in the same reaction show that the levels of transcription seen are additive, thus arguing against the possibility of some trans-acting transcription inhibitor in the $p\varkappa/E\varkappa$ experiments. This result can be observed with both the $\varkappa$ enhancer and the heavy-chain enhancer. Finally, this repressive effect of the enhancer is not observed in transcription experiments done under identical conditions in a HeLa extract. All the effects mentioned above have been

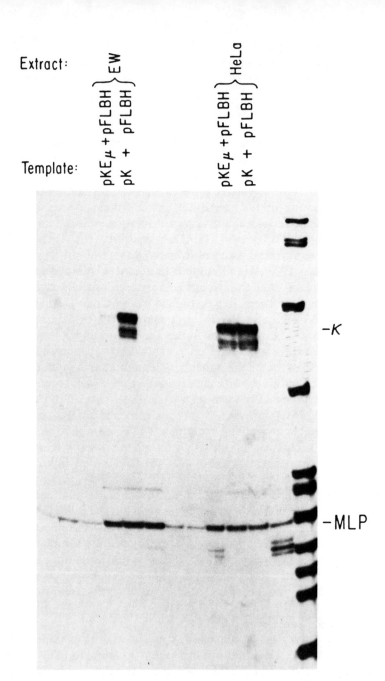

Extract:

$pKE\mu + pFLBH$ ⎱ EW
$pK + pFLBH$ ⎰

$pKE\mu + pFLBH$ ⎱ HeLa
$pK + pFLBH$ ⎰

Template:

$-\kappa$

$-MLP$

Closed circular templates

Figure 1 *See facing page for legend.*

94

reproduced in two independent extracts from the same cell line.

The promoter region of each x gene examined to date contains the conserved octamer ATTGCAT present 65 bp upstream of the cap site (Parslow et al. 1984). The region containing this sequence has been shown to be important for x gene transcription by Falkner and Zachau (1984) and Bergman et al. (1984). We have examined, in vitro, the transcription of a 5' deletion mutation that does not contain this conserved octamer and has been shown to be inactive in transfection experiments (Bergman et al. 1984). pΔx and pΔx/Ex are templates with and without the μ enhancer, and Figure 2 shows the results of transcription reactions carried out in B-lymphoma extracts in the presence of PEG after the preincubation pulse-chase protocol described above. On linear templates or closed circular templates, there is a considerable decrease in the extent of transcription from the deleted template. In constrast to the results obtained with the lymphocyte-specific enhancers, however, the effect of the upstream sequence is seen (albeit to a somewhat lesser extent) in a HeLa extract, as well. This is in agreement with the observation that deletion of the octamer drastically reduced transcription from the x promoter when transfected into non-B lymphoid cells, as well (Falkner et al. 1984; Y. Bergman et al., unpubl.).

Our results can be summarized as follows:

1. We have developed a reconstituted transcription system from the mouse myeloma MPC-11 for the analysis of B-cell-specific genes.
2. We have obtained transcriptionally competent whole cell extracts from two different B-lymphoma lines.
3. In the B-lymphoma extract under certain transcription conditions, templates containing either the heavy-chain enhancer or the x enhancer are specifically suppressed. This suppression is not observed in HeLa extracts.

Figure 1 (*see facing page*) Transcription of closed circular templates in B-cell extracts and HeLa extracts. Each reaction contains, in addition to the x gene, a plasmid containing the major late promoter of adenovirus (pFLBH) for use as an internal control. The reactions were carried out using a preincubation pulse-chase protocol as described in the text. x- and adenovirus-specific RNAs were selected by hybridization to M13 probes containing the respective cap sites and then digested with RNase T1 to generate the specific bands shown in the figure. Transcripts originating from the x gene and the major late promoter are labeled x and MLP, respectively. There is a strong repression of in vitro transcription from enhancer-containing templates, and the effect is specific to the B-cell extracts only.

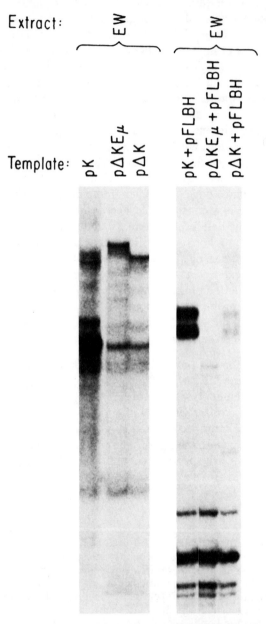

Extract: EW   EW

Template: pK   pΔKEμ   pΔK   pK+pFLBH   pΔKEμ+pFLBH   pΔK+pFLBH

Linear template   Closed circular template

Figure 2 *See facing page for legend.*

96

4. A mutant of the x promoter that deletes the conserved octamer is transcribed 10-fold less efficiently than the wild-type promoter. This effect is, however, not tissue specific in transfection experiments and can be seen in B-lymphoma extracts as well as HeLa extracts.

DISCUSSION

The specific repression observed from enhancer-containing templates in B-cell extracts is intriguing, although at present we do not have any good evidence to correlate this effect with the observed up-regulation of transcription in vivo. However, some speculations on the origin of this phenomenon are indicated. One possibility is that activation of an enhancer to increase transcription is a multicomponent phenomenon in which there are tissue-specific as well as nonspecific factors that interact (e.g., Mercola et al. 1985). Maybe in the transcription reactions described, the complete complement of factors are not binding (however, the tissue-specific ones are), thus giving the abnormal phenotype observed. A second possibility is that this is an artifactual effect of binding a tissue-specific activator. Such an artifact could arise either by physical blockage of the template for transcription by RNA polymerase II due to the binding of the factor or by sequestering of the templates in a way so as to make them unavailable for binding by other transcription factors. A third possibility is that our view of the mechanism of enhancer function has been biased by the early proposal that an enhancer might serve as an entry site for polymerase (Moreau et al. 1981; Wasylyk et al. 1983), which has the direct, experimentally unverifiable (in vivo) consequence that its effect must be in the rate of polymerase loading. We now know that some protein is bound in a tissue-specific manner to the μ enhancer (Church et al. 1985; Ephrussi et al. 1985). If the role of this protein in vivo is not to serve as a binding site for polymerase II, but rather, for example, to de-

Figure 2 (*see facing page*)   Effect of "upstream" sequences on immunoglobulin transcription in vitro. The deletion mutant pΔx is truncated at -50 relative to the cap site. The 5' end was converted to a *Hin*dIII site, and the gene until the *Bgl*II site was cloned into the *Hin*dIII-*Bam*HI sites of pUC-13. pΔxEμ contains the 700-bp heavy-chain enhancer placed in the upstream *Hin*dIII site. These plasmids were linearized with *Sac*I (which cuts in the pUC-13 polylinker) for runoff transcription analysis. The left three lanes show the use of linearized templates to observe the effects of the upstream deletion. The right three lanes utilize closed circular templates with the major late promoter (MLP) of adenovirus as an internal control. The expected transcription products are labeled x (for the x gene) and MLP.

fine nuclear localization or to change chromatin structure, then its effect in vitro need not necessarily be seen as an increase in the number of polymerase molecules accurately initiating transcription during the pulse. Finally, it has recently been shown that the SV40 and polyoma enhancers may behave like repressors of transcription under certain conditions (Borelli et al. 1984). It is possible that our in vitro assay is generating a set of conditions that is making the immunoglobulin enhancers act as repressors of transcription. We are in the process of further characterizing this effect with the aim of being able to assess unequivocally its physiological relevance by (1) utilizing well-characterized mutant enhancer sequences in in vitro transcription reaction and by (2) partial purification of the putative factor or factors that confer this unexpected effect.

ACKNOWLEDGMENTS

R.S. is a fellow of the Damon Runyon–Walter Winchell Cancer Fund. This work was supported by a grant from the American Cancer Society.

REFERENCES

Banerji, J., L. Olson, and W. Schaffner. 1983. A lymphocyte specific cellular enhancer is located downstream of the joining region in immunoglobulin heavy chain genes. *Cell* **33**: 729.

Bergman, Y., D. Rice, R. Grosschedl, and D. Baltimore. 1984. Two regulatory elements for immunoglobulin kappa light chain gene expression. *Proc. Natl. Acad. Sci.* **81**: 7041.

Borelli, E., R. Hen, and P. Chambon. 1984. Adenovirus-2 E1A products repress enhancer-induced stimulation of transcription. *Nature* **312**: 608.

Church, G.M., A. Ephrussi, W. Gilbert, and S. Tonegawa. 1985. Cell type-specific contacts to immunoglobulin enhancers in nuclei. *Nature* **301**: 798.

Davison, B.L., J.M. Egly, E.R. Mulvihill, and P. Chambon. 1983. Formation of stable pre-initiation complexes between eukaryotic class B transcription factors and promoter sequences. *Nature* **301**: 680.

Dynan, W.S. and R. Tjian. 1983. The promoter specific transcription factor Sp1 binds to upstream sequences in the SV40 early promoter. *Cell* **35**: 79.

Ephrussi, A., G.M. Church, S. Tonegawa, and W. Gilbert. 1985. B lineage-specific interactions of an immunoglobulin enhancer with cellular factors in vivo. *Science* **227**: 134.

Falkner, F.G. and H.G. Zachau. 1984. Correct transcription of an immunoglobulin kappa gene requires an upstream fragment containing conserved sequence elements. *Nature* **310**: 71.

Falkner, F.G., E. Neumann, and H.G. Zachau. 1984. Tissue specificity of the initiation of immunoglobulin kappa gene transcription. *Hoppe-Seyler's Z. Physiol. Chem.* **3655**: 1331.

Gillies, S.D., S.L. Morrison, V.T. Oi, and S. Tonegawa. 1983. A tissue specific transcription enhancer element is located in the major intron of a rearranged immunoglobulin heavy chain gene. *Cell* **33**: 717.

Grosschedl, R. and M.L. Birnstiel. 1982. Delimitation of far upstream sequences required for maximal in vitro transcription of an H2A histone gene. *Proc. Natl. Acad. Sci.* **79:** 297.

Hansen, U. and P.A. Sharp. 1983. Sequence controlling in vitro transcription of SV40 promoters. *EMBO J.* **2:** 2293.

Hen, R., P. Sassone-Corsi, J. Corden, M.P. Gaub, and P. Chambon. 1982. Sequences upstream from the TATA box are required in vivo and in vitro for efficient transcription from the adenovirus serotype 2 major late promoter. *Proc. Natl. Acad. Sci.* **79:** 7132.

Jove, R. and J.L. Manley. 1984. In vitro transcription from the adenovirus 2 major late promoter utilizing templates truncated at promoter proximal sites. *J. Biol. Chem.* **259:** 8513.

Manley, J.L., A. Fire, A. Cano, P.A. Sharp, and M.L. Gefter. 1980. DNA dependent transcription of adenovirus genes in a soluble whole cell extract. *Proc. Natl. Acad. Sci.* **77:** 3855.

Mercola, M., J. Goverman, C. Mirell, and K. Calame. 1985. Immunoglobulin heavy chain enhancer requires one or more tissue-specific factors. *Science* **227:** 266.

Moreau, P., R. Hen, B. Wasylyk, R.D. Everett, M.P. Gaub, and P. Chambon. 1981. The SV40 72 base pair repeat has a striking effect on gene expression in both SV40 and other chimeric recombinants. *Nucleic Acids Res.* **9:** 6047.

Parker, C.S. and J. Topol. 1984a. A *Drosophila* RNA polymerase II transcription factor contains a promoter-region-specific DNA-binding activity. *Cell* **36:** 357.

―――. 1984b. A *Drosophila* RNA polymerase II transcription factor binds to the regulatory site of an hsp70 gene. *Cell* **37:** 273.

Parslow, T.G., D.L. Blair, W.J. Murphy, and D.K. Granner. 1984. Structure of the 5' ends of immunoglobulin genes: A novel conserved sequence. *Proc. Natl. Acad. Sci.* **81:** 2650.

Picard, D. and W. Schaffner. 1984. A lymphocyte specific enhancer in the mouse immunoglobulin kappa gene. *Nature* **307:** 80.

Queen, C. and D. Baltimore. 1983. Immunoglobulin gene transcription is activated by downstream sequence elements. *Cell* **33:** 741.

Sassone-Corsi, P., J. Dougherty, B. Wasylyk, and P. Chambon. 1984. Stimulation of in vitro transcription from heterologous promoters by the simian virus 40 enhancer. *Proc. Natl. Acad. Sci.* **81:** 308.

Sergeant, A., D. Bohmann, H. Zentgraf, H. Weiher, and W. Keller. 1984. A transcription enhancer acts in vitro over distances of hundreds of basepairs on both circular and linear templates but not on chromatin reconstituted DNA. *J. Mol. Biol.* **180:** 577.

Tsuda, M. and Y. Suzuki. 1981. Faithful transcription initiation of fibroin gene in a homologous cell-free system reveals an enhancing effect of 5' flanking sequence far upstream. *Cell* **27:** 175.

Wasylyk, B., C. Wasylyk, P. Augereau, and P. Chambon. 1983. The SV40 72 base pair repeat preferentially potentiates transcription starting from proximal natural or substitute promoter elements. *Cell* **32:** 503.

Weil, P.A., D.S. Luse, J. Segall, and R.G. Roeder. 1979. Selective and accurate initiation of transcription at the Ad2 major late promoter in a soluble system dependent on purified RNA polymerase I and DNA. *Cell* **18:** 469.

Wildeman, A.G., P. Sassone-Corsi, T. Grundstrom, M. Zenke, and P. Chambon. 1984. Stimulation of in vitro transcription from the SV40 early promoter by the enhancer involves a specific trans-acting factor. *EMBO J.* **3:** 3129.

Cell-specific Factors Required for Transcriptional Enhancement In Vitro

H.R. Schöler, U. Schlokat, and P. Gruss

Zentrum für Molekulare Biologie der Universität Heidelberg
Im Neuenheimer Feld 364, D-6900 Heidelberg
Federal Republic of Germany

Enhancer elements drastically increase the transcriptional activity of many promoters relatively independently of orientation and distance with respect to the coding region (for review, see Gluzman and Shenk 1983; Khoury and Gruss 1983). The cell-specific activity of some enhancers is of particular interest, including that of the immunoglobulin enhancers (Banerji et al. 1983; Gillies et al. 1983; Queen and Baltimore 1983), insulin and chymotrypsin enhancers (Walker et al. 1983; W. Rutter, pers. comm.), the elastase I enhancer (Swift et al. 1984), and viral enhancers such as the lymphotropic papovavirus (LPV) enhancer (Mosthaf et al. 1985). Since cellular factors are required for enhancer activity in vivo (Schöler and Gruss 1984) and in vitro (Sassone-Corsi et al. 1985), it is reasonable to speculate that the cell-specific activity of many enhancers is mediated by their interaction with specific cellular factors. Using two different experimental approaches, we have attempted to define these cellular factors with the help of in vitro assays.

RESULTS AND DISCUSSION

Cell-free systems that accurately and specifically transcribe exogenous DNA templates for polymerase II have been described previously (Weil et al. 1979; Manley et al. 1980). Most experiments were performed using human cells (HeLa, KB) growing in suspension to prepare the extracts. With the help of these cell-free systems, it has recently been demonstrated that the presence of SV40 enhancer sequences at or near the 5' end of different promoters leads to a 5-fold to 10-fold increase of transcriptional activity of exogenous DNA (Sassone-Corsi et al. 1984; Sergeant et al. 1984). Although the majority of transcripts start at the correct position, several classes of RNA were transcribed from positions not recognized in in vivo experiments. The presence of these RNA classes, which most likely start unspecifically, can be greatly reduced if, instead of a whole cell extract, a nuclear extract is prepared (Wildeman et al. 1984).

In our initial experiments we used such a nuclear extract prepared from HeLa cells to test for transcriptional enhancement of several different enhancer elements. All constructions used had been tested previously in in vivo experiments (Mosthaf et al. 1985) and are outlined schematically at the bottom of Figure 1. In brief, we employed four constructions that all carry identical SV40 promoter sequences (21-bp repeats, TATA box) in front of a prokaryotic gene (cat) and either one of three different enhancer elements (SV40, IgCμ, LPV) or none upstream of it. In in vivo experiments this general setup allowed quantification of both RNA and protein. However, the amount of RNA produced was determined only for the in vitro experiments. The probe employed for S1 nuclease analysis was a terminally labeled fragment from BglII-EcoRI deriving from plasmid pIgCAT. In our S1 nuclease experiments, RNA produced from the SV40 promoter (early-early [EE] start sites) was expected to be around 310 nucleotides in length, whereas the RNA transcribed from the test plasmid should yield a fragment of 250 nucleotides.

Not surprisingly, and in good agreement with previous in vivo data (Mosthaf et al. 1985), a stimulation of transcription was observed in HeLa cell extracts only when SV40 enhancer sequences were present 5' of the SV40 promoter (up to 10-fold varying from extract to extract). No increase in transcriptional activity was observed if the LPV or mouse IgCμ enhancer was present on the template DNA (data not presented). This result could be explained if either a factor required for LPV and IgCμ enhancer activity is missing or, likewise, if factors negatively interacting with LPV and IgCμ enhancer sequences are present in HeLa cell extracts. Thus, to test whether these enhancers exhibit in vitro activity in cells in which they show activity in vivo, we subsequently prepared nuclear extracts from established human lymphoid cell lines. Specifically, we

Figure 1 (*see facing page*)   In vitro transcription using nuclear extracts prepared from MOLT-4. Nuclear extracts prepared from a human T-cell line were prepared as described previously for HeLa cells (Wildeman et al. 1984). This extract was used for in vitro transcription of pA10CAT2, pSV2CAT, pIgCAT, and pLPVCAT. All plasmids used are virtually identical except for the presence or absence of enhancer sequences. pSV2CAT denotes the presence of the SV40 enhancer; pIgCAT denotes the presence of the mouse IgCμ enhancer; and pLPVCAT denotes the presence of the enhancer derived from a lymphotropic papovavirus. All enhancer elements are located 5' in front of SV40 promoter sequences (21-bp repeat, TATA box) and were tested previously for their in vivo activity.

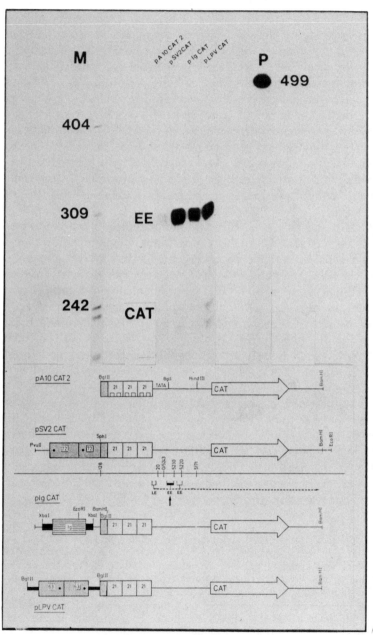

Figure 1 *See facing page for legend.*

103

used BJA-B, a B-cell line (EBV-negative), and MOLT-4, a T-cell line deriving from a patient with acute lymphoblastic leukemia of the T-cell type. Using the MOLT-4 extract, specific transcripts starting at the SV40 EE start sites (Tooze 1981) were produced by all four templates (Fig. 1). However, differences in the amount of RNA synthesized are clearly revealed. A template lacking all enhancer sequences (pA10CAT2) yields the lowest signal. An approximately 10-fold increase can be observed if either the SV40, $IgC\mu$, or LPV enhancer is present on the template. Similar results were obtained with the BJA-B extract (data not shown). Thus, in vivo cell-specific activity of these enhancers can be reproduced in vitro. Again, we cannot distinguish yet between negative or positive control mechanisms. If the enhancers active in lymphoid cells are activated by binding of a positively acting component, one would expect to discover factors specifically interacting with these enhancer sequences.

To this end we have utilized the nitrocellulose filter–binding assay (Riggs et al. 1970) to detect specific binding factors present in BJA-B cells. With this assay no specific binding to the $IgC\mu$ enhancer could be detected in the crude nuclear extracts used for transcriptional analysis. Since this failure could be due to the abundance of unspecific DNA-binding proteins masking the specific binding activity, we devised a fractionation scheme outlined in Figure 2 (top). Nuclei of BJA-B cells were extracted with NaCl (Nowock and Sippel 1982) at a concentration that preferentially allowed elution of DNA-binding proteins; in contrast, elution of histones in particular was by and large prevented.

To remove possible contaminating cellular chromatin, the eluate was applied to a DEAE column. Under the conditions used, chromatin is retained on the column whereas the vast majority of soluble factors appear in the flow-through. Subsequently, the flow-through was fractionated by stepwise NaCl elution after application to a heparin column. All fractions were tested for DNA binding. Only one fraction revealed specificity in binding (D_4).

Figure 2 (*see facing page*) Fractionation scheme and nitrocellulose filter–binding assay. Human BJA-B cells were grown in suspension and fractionated as outlined at *top*. The last column indicated (heparin) was eluted by increasing the NaCl concentration stepwise. Fraction D_4 showed specificity in a subsequent nitrocellulose filter–binding assay. In this assay we used plasmid $pC\mu5'$-CATa, which upon cleavage with *Cla*I and terminal labeling using $[\alpha\text{-}^{32}P]ATP$ carried the mouse $IgC\mu$ enhancer in the larger fragment (A). As shown at lower right, this fragment is preferentially retained in the nitrocellulose filter–binding assay using the D_4 fraction.

FRACTIONATION OF DNA BINDING PROTEINS FROM BJA-B CELLS

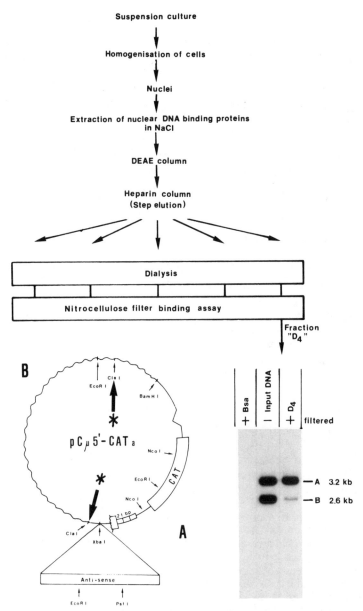

Figure 2 *See facing page for legend.*

As shown in Figure 2 (bottom), a terminally labeled fragment containing the mouse IgCμ enhancer sequences was specifically retained on the filter. Subsequent analysis demonstrated that the binding is due to the presence of the 1-kb Cμ enhancer (data not shown). In contrast, no binding activity to the SV40 21-bp repeat promoter region (Gidoni et al. 1984) could be detected in this fraction under the assay conditions used. So far, mixing the D₄ fraction with HeLa cell nuclear extract has not resulted in a change in the transcriptional activation pattern in HeLa cells. Therefore, no evidence is available yet demonstrating functional significance of the factors present in the D₄ fraction that binds specifically to Ig enhancer sequences.

Currently, we are in the process of fractionating the HeLa cell and BJA-B extracts. Crosswise complementation also using the D₄ fraction should help to identify the components required for the enhancement effect in vitro. This understanding seems to be a prerequisite for the elucidation of the mechanism of enhancer function.

ACKNOWLEDGMENTS
The isolation of enhancer-binding proteins represents part of a collaborative effort with J. Nowock and A. Sippel. This was was supported by BMFT grant BCT 0364/1.

REFERENCES

Banerji, I., L. Olson, and W. Schaffner. 1983. A lymphocyte-specific cellular enhancer is located downstream of the joining region in immunoglobulin heavy chain genes. *Cell* **33:** 729.

Gidoni, D., W.S. Dynan, and R. Tjian. 1984. Multiple specific contacts between a mammalian transcription factor and its cognate promoters. *Nature* **312:** 409.

Gillies, S.D., S.L. Morrison, V.T. Oi, and S. Tonegawa. 1983. A tissue-specific transcription enhancer element is located in the major intron of a rearranged immunoglobulin heavy chain gene. *Cell* **33:** 717.

Gluzman, Y. and T. Shenk, eds. 1983. *Current communications in molecular biology: Enhancers and eukaryotic gene expression.* Cold Spring Harbor Laboratory, Cold Spring Harbor, New York.

Khoury, G. and P. Gruss. 1983. Enhancer elements. *Cell* **33:** 313.

Manley, J.L., A. Fire, A. Cano, P.A. Sharp, and M.L. Gefter. 1980. DNA-dependent transcription of adenovirus genes in a soluble whole-cell extract. *Proc. Natl. Acad. Sci.* **77:** 3855.

Mosthaf, L., M. Pawlita, and P. Gruss. 1985. A viral enhancer element specifically active in human hematopoietic cells. *Nature* (in press).

Nowock, J. and A.E. Sippel. 1982. Specific protein-DNA interaction at four sites flanking the chicken lysozyme gene. *Cell* **30:** 607.

Queen, C. and D. Baltimore. 1983. Immunoglobulin gene transcription is activated by downstream sequence elements. *Cell* **33:** 741.

Riggs, A.D., H. Suzuki, and S. Bourgois. 1970. Lac repressor operator interaction. I. Equilibrium studies. *J. Mol. Biol.* **48:** 67.

Sassone-Corsi, P., A. Wildeman, and P. Chambon. 1985. A trans-acting factor is responsible for the simian virus 40 enhancer activity in vitro. *Nature* **313:** 458.

Sassone-Corsi, P., J.P. Dougherty, B. Wasylyk, and P. Chambon. 1984. Stimulation of in vitro transcription from heterologous promoters by the simian virus 40 enhancer. *Proc. Natl. Acad. Sci.* **81:** 308.

Schöler, H.R. and P. Gruss. 1984. Specific interaction between enhancer-containing molecules and cellular components. *Cell* **36:** 403.

Sergeant, A., D. Bohmann, H. Zentgraf, H. Weiher, and W. Keller. 1984. A transcription enhancer acts in vitro over distances of hundreds of basepairs on both circular and linear templates but not on chromatin-reconstituted DNA. *J. Mol. Biol.* **180:** 577.

Swift, G.H., R.E. Hammer, R.J. MacDonald, and R.L. Brinster. 1984. Tissue-specific expression of the rat pancreatic elastase I gene in transgenic mice. *Cell* **38:** 639.

Tooze, J., ed. 1981. *Molecular biology of tumor viruses*, 2nd edition, revised: *DNA tumor viruses*. Cold Spring Harbor Laboratory, Cold Spring Harbor, New York.

Walker, M.D., T. Edlund, A.M. Boulet, and W.J. Rutter. 1983. Cell specific expression controlled by the 5'-flanking region of insulin and chymotrypsin genes. *Nature* **306:** 557.

Weil, P.A., D.S. Luse, J. Segall, and R.G. Roeder. 1979. Selective and accurate initiation of transcription at the Ad2 major late promoter in a soluble system dependent on purified RNA polymerase II and DNA. *Cell* **18:** 469.

Wildeman, A., P. Sassone-Corsi, T. Grundström, M. Zenke, and P. Chambon. 1984. Stimulation of in vitro transcription from the SV40 early promoter by the enhancer involves a specific trans-acting factor. *EMBO J.* **13:** 3129.

Gene-specific RNA Polymerase II Transcription Factors

K.A. Jones and R. Tjian

Department of Biochemistry, University of California at Berkeley
Berkeley, California 94720

Accurate expression of protein-coding genes in eukaryotic systems requires promoter elements located upstream of the gene that determine the frequency of transcriptional initiation, position the RNA start site, and specify the DNA strand to be transcribed. Extensively purified RNA polymerase II will not recognize its cognate promoters (Weil et al. 1979) but, rather, requires auxiliary factors to accurately initiate transcription at a given promoter in vitro (Matsui et al. 1980; Samuels et al. 1982). In addition, optimal RNA synthesis in vivo requires *cis*-acting enhancer elements (c.f. Benoist and Chambon 1981) or *trans*-acting inducer proteins (c.f. Nevins 1981) that can act in concert with the promoter. Clearly, a complete understanding of the events involved in transcriptional initiation and regulation will ultimately require a description of the mechanism of promoter recognition by RNA polymerase II.

Initial comparisons of individual promoter sequences focused on a TATA or AT-rich element located 20–30 bp upstream of the RNA start site and a "CAAT" sequence located 50 bp further upstream. However, it soon became apparent that these promoter elements were not strictly conserved among all RNA polymerase II genes and that detailed studies of many individual promoter elements would be required.

Two promoters that have been extensively characterized are those of the SV40 early and the herpes simplex virus (HSV) thymidine kinase (*tk*) genes. These two genes contain transcriptional control regions located approximately 50–100 bp upstream of the RNA start site that are required for full promoter activity in vivo (Benoist and Chambon 1981; McKnight 1982). Both upstream control regions can function in an orientation-independent manner and contain repeats of a hexanucleotide sequence, GGGCGG, known as a GC box. An extract fractionation-reconstitution approach (described below) that was designed to characterize the proteins involved in the initiation of RNA synthesis in vitro has proved useful in understanding the events leading to RNA polymerase II recognition of the SV40 and *tk* promoters.

The Sp1 Transcription Factor

The SV40 upstream control region is characterized by six tandem GC-box repeats. Through fractionation of a HeLa whole cell extract, Dynan and Tjian (1983a,b) identified a cellular protein, Sp1, that binds to the SV40 early promoter and activates transcription in a reconstituted reaction with partially purified RNA polymerase II and other general transcription factors. Gidoni et al. (1984) subsequently showed that Sp1 interacts specifically with the GC-box sequences, and a detailed analysis of the interaction with clustered base-substitution promoter mutants suggests that the SV40 upstream control region contains at least three tandem Sp1-binding sites (D. Gidoni et al., unpubl.). Although Sp1 was originally identified as an SV40-specific transcription factor that is not required for all RNA polymerase II promoters, it seemed likely that this factor would also mediate polymerase recognition of a class of cellular genes and perhaps other viral genes, as well. Therefore it was of interest to identify other genes that respond to Sp1, both to elucidate the role of Sp1 in the cell and to analyze its interaction with different promoter sequences.

HSV Immediate-early Genes

The HSV genome is expressed in three temporally regulated stages, corresponding with the transcription of immediate-early (IE), early, and late genes (Honess and Roizman 1974). All five IE genes contain multiple GC boxes in the promoter and upstream regulatory regions (Post et al. 1981; Cordingley et al. 1983; Lang et al. 1984; Preston et al. 1984). In the in vitro reconstituted system, transcription of the IE-3 (ICP-4) and IE-4,5 genes was found to be dramatically responsive to Sp1, and multiple-unit Sp1-binding sites were mapped in the promoter and adjacent regulatory regions of both genes in DNaseI footprinting experiments (Jones and Tjian 1985). The individual Sp1-binding sites protect approximately 18–20 bp each and are centered on a single GC box, and binding was independent of the GC-box orientation. Some variation in the relative affinity of these sites for Sp1 was observed, and three GC boxes did not appear to be recognized at all by Sp1, indicating that the recognition sequence is larger than the hexanucleotide sequence per se. Comparison of the HSV GC boxes with those of the SV40 promoter suggests a minimal recognition sequence of $^T_GGGGCGG^A_GGC$; however, a larger number of individual Sp1-binding sites must be evaluated before a recognition sequence can be identified with certainty.

The HSV *tk* Gene

The *tk* gene was an excellent candidate for in vitro analysis because the properties of the upstream control region have been characterized extensively in vivo through the analysis of linker-scanner and single-base-substitution promoter mutants (McKnight 1982; McKnight and Kingsbury 1982; McKnight et al. 1984). These studies revealed that *tk* transcription requires the two inverted GC-box repeats of the upstream control region. In addition, the promoter was found to contain a distinct set of mutationally sensitive sequences, suggesting that *tk* transcription requires a factor other than Sp1 that could either act alone to recognize the complete upstream control region or could act in conjunction with Sp1. A detailed characterization of the *tk* promoter in vitro, using a combination of the mutational analysis and the extract fractionation approaches, revealed that the *tk* promoter is recognized by two chromatographically separable types of transcription factors. These two factors bind with different affinities to three sites in the upstream control region, and linker-scanner mutations that affect binding also affect transcription (K. Jones et al., in prep.). Binding to the TATA-proximal site is extremely weak and difficult to characterize, whereas the interactions at the other sites are moderate in affinity. The transcription factor that binds to the second distal GC box has been shown to copurify extensively with Sp1. The other moderate-affinity binding site contains an inverted CAAT sequence, which is required for binding, and fractions containing this factor also recognize the CAAT sequence in the β-globin and MO-MSV promoters (K. Jones and R.T. Tjian, unpubl.). Thus, this factor may represent a CAAT-binding transcription factor (CTF) that can activate transcription in an orientation-independent fashion, like Sp1. These results indicate that Sp1 and CTF act in conjunction to mediate RNA polymerase II recognition of the *tk* upstream control region.

Other Sp1-responsive Promoters

Parallel investigations of both binding and transcriptional activation suggest that the mouse dihydrofolate reductase and human metallothionein-IA genes also belong to the Sp1-responsive gene class (W. Dynan et al.; W. Lee et al.; both unpubl.).

Future Directions

Although the criteria of using extensively purified protein and demonstrating specific competition for binding with SV40 promoter

DNA are strongly suggestive, it is clear that proof that each of these promoters is in fact recognized by Sp1 and not by some other transcription factor with similar binding properties awaits the purification of Sp1 to homogeneity. Furthermore, the complete characterization of *tk* transcription will also require the purification of CTF and therefore these goals are a major focus of research in the lab at the present time.

ACKNOWLEDGMENTS

We thank Jas Lang and Neil Wilkie for HSV IE promoter plasmids, and Steve McKnight for HSV *tk* promoter mutants.

REFERENCES

Benoist, C. and P. Chambon. 1981. In vivo sequence requirements for the SV40 early promoter region. *Nature* **290**: 304.

Cordingley, M.G., M.E.M. Campbell, and C.M. Preston. 1983. Functional analysis of a herpes simplex virus type 1 promoter: Identification of far-upstream regulatory sequences. *Nucleic Acids Res.* **11**: 2347.

Dynan, W.S. and R. Tjian. 1983a. Isolation of transcription factors that discriminate between different promoters recognized by RNA polymerase II. *Cell* **32**: 669.

———. 1983b. The promoter-specific transcription factor Sp1 binds to upstream sequences in the SV40 early promoter. *Cell* **35**: 79.

Gidoni, D., W.S. Dynan, and R. Tjian. 1984. Multiple specific contacts between a mammalian transcription factor and its cognate promoter. *Nature* **312**: 409.

Honess, R.W. and B. Roizman. 1974. Regulation of herpesvirus macromolecular synthesis. I. Cascade regulation of the synthesis of three groups of viral proteins. *J. Virol.* **14**: 8.

Jones, K.A. and R. Tjian. 1985. Sp1 binds to promoter sequences and activates HSV "immediate-early" gene transcription in vitro. *Nature* (in press).

Lang, J.C., D.A. Spandidos, and N.M. Wilkie. 1984. Transcriptional regulation of a herpes simplex virus immediate early gene is mediated through an enhancer-type sequence. *EMBO J.* **3**: 389.

Matsui, T., J. Segall, P.A. Weil, and R.G. Roeder. 1980. Multiple factors required for accurate initiation of transcription by purified RNA polymerase II. *J. Biol. Chem.* **255**: 11992.

McKnight, S.L. 1982. Functional relationships between the transcriptional control signals of the thymidine kinase gene of herpes simplex virus. *Cell* **31**: 355.

McKnight, S.L. and R. Kingsbury. 1982. Transcriptional control signals of a eukaryotic protein-coding gene. *Science* **217**: 316.

McKnight, S.L., R. Kingsbury, A. Spence, and M. Smith. 1984. The distal transcription signals of the herpesvirus *tk* gene share a common hexanucleotide control sequence. *Cell* **37**: 253.

Nevins, J.R. 1981. Mechanism of activation of early viral transcription by the adenovirus E1A gene product. *Cell* **26**: 213.

Post, L.E., S. Mackem, and B. Roizman. 1981. Regulation of α-genes of herpes simplex virus: Expression of chimeric genes produced by fusion of thymidine kinase with α-gene promoters. *Cell* **24:** 555.

Preston, C.S., M.G. Cordingley, and N.D. Stow. 1984. Analysis of DNA sequences which regulate the transcription of a herpex simplex virus immediate early gene. *J. Virol.* **50:** 708.

Samuels, M., A. Fire, and P. Sharp. 1982. Separation and characterization of factors mediating accurate transcription by RNA polymerase II. *J. Biol. Chem.* **257:** 14419.

Weil, P.A., D.S. Luse, J. Segall, and R.G. Roeder. 1979. Selective and accurate initiation of transcription at the Ad2 major late promoter in a soluble system dependent on purified RNA polymerase II and DNA. *Cell* **18:** 469.

Adenovirus E1A Stimulates Transcription of Class-III Genes by Increasing the Activity of Transcription Factor IIIC

S.K. Yoshinaga, N. Dean, L.T. Feldman,
R.G. Gaynor, and A.J. Berk
Molecular Biology Institute, University of California at Los Angeles
Los Angeles, California 90024

Proteins encoded in early region 1A (E1A) of group-C adenoviruses (Ad2 and Ad5) stimulate transcription from six viral promoters active during the early phase of infection (Berk et al. 1979; Jones and Shenk 1979a; Nevins 1981). Several lines of evidence indicate that this an indirect effect of E1A proteins. In a mutant, *dl*312 (Jones and Shenk 1979b), in which the E1A region is completely deleted, transcription from the other early regions occurs in infected HeLa cells but is greatly delayed compared with transcription of these regions in HeLa cells infected with wild-type virus (Nevins 1981; Gaynor and Berk 1983). E1A functions can also stimulate transcription of nonadenovirus genes in transient transfection assays, including the human and rabbit β-globin genes (Green et al. 1983; Svensson and Akusjarvi 1984), the rat preproinsulin I gene (Gaynor et al. 1984), an early SV40 promoter–α-globin hybrid gene from which both copies of the SV40 72-bp enhancer region have been deleted (Treisman et al. 1983), and the herpes simplex virus type 1 (HSV-1) glycoprotein D gene (an early or β-class HSV-1 gene; Everett and Dunlop 1984). Also, the pseudorabies virus (a herpes virus) immediate-early (IE) protein can complement adenovirus E1A mutations despite the fact that the adenoviruses and pseudorabies virus do not share significant homologies (Feldman et al. 1982). Like the adenovirus E1A proteins, herpes virus IE proteins also stimulate transcription of transfected β-globin genes (Everett 1983; Green et al. 1983).

We find that expression of adenovirus E1A functions (perhaps in conjunction with other adenovirus functions) and pseudorabies virus IE protein stimulate transcription of transfected class-III genes

(genes transcribed by RNA polymerase III) as well as transfected class-II genes.

Plasmid clones of adenovirus VA-I and *Drosophila* tRNA arginine (Silverman et al. 1979) genes were transfected into HeLa cells, 293 cells (human embryonic kidney cells constitutively expressing Ad5 E1A and E1B proteins), 143 cells, PR143 cells (143 cells transformed with and constitutively expressing the pseudorabies virus IE protein), and Ad2-infected HeLa cells. At least 10-fold higher concentrations of these class-III RNAs were found in cytoplasmic RNA isolated 48 hr posttransfection from 293, PR143, and Ad2-infected HeLa cells than from HeLa or 143 cells. Control experiments showed that this was not due to differences in transfection efficiency.

In vitro transcription of VA-I with S100 extracts showed that extracts of 293 cells were 5- to 10-fold more active than extracts of HeLa cells. Extracts of HeLa cells infected with Ad5 and held in the early phase of infection by treatment with the DNA synthesis inhibitor cytosine arabinoside were 5-fold more active than extracts from HeLa cells infected with the E1A deletion mutant *dl*312. Fractionation of these extracts indicated that the increased activity was not due to E1A protein present in the extracts. Rather, the results indicated an increase in the activity of the rate-limiting class-III transcription factor, TFIIIC. These results suggest that the increased expression of transfected class-III genes in cells expressing E1A protein results from an increase in the in vivo activity of TFIIIC. These findings raise the possibility that E1A protein stimulates the transcription of class-II genes by causing an increase in the activity of key class-II transcription factors.

ACKNOWLEDGMENTS

This work was supported by grants CA-25235 and CA-32737 from the National Cancer Institute.

REFERENCES

Berk, A.J., F. Lee, T. Harrison, J. Williams, and P.A. Sharp. 1979. Pre-early adenovirus 5 gene products regulate transcription and processing of early viral messenger RNAs. *Cell* **17:** 935.

Everett, R.D. 1983. DNA sequence elements required for regulated expression of the HSV-1 glycoprotein D gene lie within 83 bp of the RNA cap sites. *Nucleic Acids Res.* **11:** 6647.

Everett, R.D. and M. Dunlop. 1984. *Trans* activation of plasmid-borne promoters by adenovirus and several herpes group viruses. *Nucleic Acids Res.* **12:** 5969.

Feldman, L.T., M.J. Imperiale, and J.R. Nevins. 1982. Activation of early adenovirus transcription by the herpes immediate early gene: Evidence for a common cellular control factor. *Proc. Natl. Acad. Sci.* **79:** 4952.

Gaynor, R.B. and A.J. Berk. 1983. *Cis*-acting induction of adenovirus transcription. *Cell* **33:** 683.

Gaynor, R.B., D. Hillman, and A.J. Berk. 1984. Adenovirus E1A protein activates transcription of a non-viral gene which is infected or transfected into mammalian cells. *Proc. Natl. Acad. Sci.* **81:** 1193.

Green, M.R., R. Treisman, and T. Maniatis. 1983. Transcriptional activation of cloned human β-globin genes by viral immediate-early gene products. *Cell* **35:** 137.

Jones, N. and T. Shenk. 1979a. An adenovirus type 5 early gene function regulates expression of other early viral genes. *Proc. Natl. Acad. Sci.* **76:** 3665.

———. 1979b. Isolation of adenovirus type 5 host range mutants defective for transformation of rat embryo cells. *Cell* **17:** 683.

Nevins, J.R. 1981. Mechanism of activation of early transcription by the adenovirus E1A gene product. *Cell* **26:** 213.

Silverman, S.D., D. Schmidt, D. Soll, and B. Hovermann. 1979. The nucleotide sequence of a cloned *Drosophila* arginine tRNA gene and its *in vitro* transcription in *Xenopus* terminal vesicle extracts. *J. Biol. Chem.* **254:** 10290.

Svensson, C. and G. Akusjarvi. 1984. Adenovirus 2 early region IA stimulates expression of both viral and cellular genes. *EMBO J.* **3:** 789.

Treisman, R., M.R. Green, and T. Maniatis. 1983. *Cis*- and *trans*-activation of globin gene transcription in transient assays. *Proc. Natl. Acad. Sci.* **80:** 7428.

Studies on the Promoter of the Herpes Virus Thymidine Kinase Gene

S. McKnight, R. Kingsbury, S. Eisenberg, F. Tufaro, S. Weinheimer, S. Triezenberg, P. Johnson, and B. Graves

Department of Embryology, Carnegie Institution of Washington
Baltimore, Maryland 21210

During its lytic infectious cycle, herpes simplex virus (HSV) exresses a thymidine kinase (tk) enzyme activity. The gene that encodes the tk enzyme has been cloned, sequenced, and shown to be capable of accurate function in a variety of in vivo and in vitro assays. tk mRNA accumulation experiments (S. McKnight, unpubl.) indicate that the tk gene is transited by RNA polymerase II only some 20 to 40 times during the prereplicative phase of the viral lytic cycle. It would appear, therefore, that the transcriptional promoter of the tk gene is rather weak. The experiments described in this report were conducted in an effort to come to grips with the issue of "promoter strength" using the tk gene as the model.

Initial experiments carried out by our group were aimed at identifying the element of the tk gene that constitutes its promoter. Early studies led to the identification of a transcriptional "control region" roughly 100 bp in length that mapped at a location immediately upstream from the tk mRNA cap site (McKnight et al. 1981). Subsequent experimentation led to the identification of three discrete domains within the tk control region (McKnight and Kingsbury 1982). One such domain houses the "TATA homology" of the tk gene. Two additional domains were delineated upstream of the TATA homology; one is positioned about 50 bp upstream from the tk mRNA cap site (termed the first distal signal, or DSI), and the other is positioned about 100 bp upstream from the cap site (termed the second distal signal, or DSII).

DSI and DSII share a common hexanucleotide sequence that occurs in inverted orientations (see Fig. 1). DSI contains the sequence 5' GGGCGG 3' between residues −55 and −50, and DSII contains the sequence 5' CCGCCC 3' between residues −103 and −98. Single-point mutations were introduced into each of these hexanucleo-

116

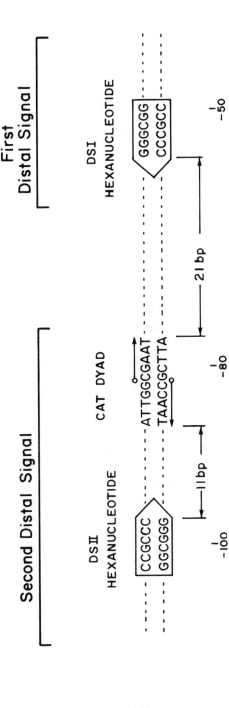

Figure 1 Diagram of distal promoter domains of the HSV *tk* gene. Shown are three promoter components that occur upstream of the HSV *tk* gene. These include the DSI hexanucleotide, which is positioned 50 bp upstream from the *tk* mRNA cap site; the CAAT ("CAT") dyad, which is positioned 80 bp upstream from the cap site; and the DSII hexanucleotide, which is positioned 100 bp upstream from the cap site. Not shown is the "TATA homology," which is positioned 30 bp upstream from the mRNA cap site.

117

tide sequences and a concordant pattern of phenotypic effects was observed. For example, C-to-G transversions at residues -103, -102, and -100 result in, respectively, mild, severe, and severe phenotypic effects on transcription. Similarly, G-to-C transversions at residues -53, -51, and -50 result in, respectively, severe, severe, and mild phenotypic effects. Since each of these transversion mutations represents the same lesion within the context of each hexanucleotide, and since the pattern of phenotypic effects was observed to be qualitatively concordant, we reasoned that the hexanucleotides might represent functional homologs and speculated that they might be recognized by the same DNA-binding protein (McKnight et al. 1983).

Although we observed a concordant pattern of phenotypic effects from the aforementioned point mutations, we also noticed that the point mutations that interrupt the hexanucleotide of DSII consistently produced a more severe reduction in expression efficiency than did the counterpart mutations of the hexanucleotide of DSI. From this discrepant feature we speculated that the binding affinity of the putative hexanucleotide-specific transcription factor might not be equivalent for both of the repeats.

Both of the speculations outlined in the preceding paragraph have been substantiated by the work of K. Jones et al. (in prep.; see also Jones and Tjian, this volume). Kathy Jones has found that the hexanucleotide repeats of the HSV *tk* genes are specifically recognized by the same transcription factor that was initially found to activate the SV40 early promoter by binding in a sequence-specific manner to the 21-bp repeats (Dynan and Tjian 1983). Indeed, the 21-bp repeats of SV40 contain six copies of the same hexanucleotide sequence that occurs in DSI and DSII of the HSV *tk* promoter. Work from the Tjian group has further shown that the apparent binding affinity of Sp1 to the hexanucleotide repeat of DSII is measurably stronger than its affinity to the repeat of DSI (see Jones and Tjian, this volume).

Our interest in pursuing the properties of the *tk* promoter stems from the early observation that a single base mutation in the hexanucleotide of DSII somehow inactivates the use of DSI. Clearly there is some sort of functional cooperativity at work in the *tk* promoter. One might speculate that binding of Sp1 to the strong binding site of DSII facilitates binding to the weaker site within DSI in the same manner that binding the *c*I repressor to the strong repressor-binding site (*orI*) of bacteriophage λ potentiates binding to the weaker *orII* site (Johnson et al. 1981). However, there are two observations that argue against a simple model of this nature. First, the two Sp1-binding sites are quite far apart (~50 bp). Second, early experimentation revealed a mutation-sensitive area between the two hexanucleotides that closely abuts the hexanucleotide of DSII (McKnight and Kingsbury 1982).

We reasoned that the mutation-sensitive area between the two Sp1-binding sites might house an independent transcriptional control signal. The primary reason for entertaining such a speculation was the occurrence of a region of dyad symmetry within the mutation-sensitive domain (see Fig. 1). To pursue this hypothesis, we created a series of 23 single-point mutations starting just downsteam from the hexanucleotide of DSII. The phenotypes of these point mutations showed that only the mutation-sensitive area occurs within the aforementioned dyad (S. Eisenberg, R. Kingsbury, and S. McKnight, in prep.).

The dyad occurs between residues − 86 and − 77, and on the noncoding strand of the *tk* gene reads 5′ ATTGGCGAAT 3′. Mutation of any residue within the left half of this dyad leads to a severe phenotypic reduction in *tk* gene function. Mutation of any residue of the right half of the dyad, with the exception of the G residue at position − 80, leads to a mild reduction in *tk* gene function. When the G residue at position − 80 is changed to C, the resulting variant is expressed some twofold better than normal. It is obvious that the dyad we have thus identified is homologous to the "CAAT-box homology" that has been identified upstream from a number of eukaryotic protein-coding genes (Efstratiadis et al. 1980).

Two surprising variations emerge from the examination of the *tk* CAAT homology. First, the critical aspect of the *tk* CAAT signal appears to occur in an inverted orientation relative to that of other eukaryotic structural genes. Second, the *tk* CAAT signal appears to consist of a dyad symmetry. Although such was not initially deduced to be the case for the CAAT homologies of the globin gene family (Efstratiadis et al. 1980), reinspection of globin gene sequences in light of our observations on the *tk* gene show that the mouse β-minor-globin gene maintains a perfect 5′ CCAAT 3′ pentanucleotide pointing away from the gene on the opposite strand of its conventional CAAT sequence. Moreover, the rabbit and human β-globin genes contain the sequence 5′ CCAAC 3′ in this position.

An independent observation that strengthens the argument that the *tk* CAAT signal is of the same nature as the prototypical CAAT homology comes from the observation of a second "up mutant" in our screen of *tk* point mutations. A C-to-G transversion at residue − 88 creates a promoter that is twofold stronger than normal. On the opposite DNA strand, this creates a G-to-C transversion that changes the inverted CAAT signal from 5′ CCAATG 3′ to 5′ CCAATC 3′. We have noticed that most CAAT signals contain a C residue at the sixth position of the sequence. Included among those that contain a C residue in the sixth position is the CAAT homology of the murine sarcoma virus (MSV) long terminal repeat (LTR). In an independent series of experiments, Barbara Graves has constructed and assayed a set of point mutants of the MSV-LTR CAAT signal (unpubl.). Dr. Graves mutated the six residues of the MSV CAAT signal (5′ CCAATC 3′) by creating exactly the same transversions as were introduced into the HSV CAAT sequence, with the exception of the final C residue (which in the case of the HSV inverted CAAT is a G residue). In scoring the effects of the MSV point mutations, Graves noted for the first five residues, a pattern of phenotypic effects very similar to those observed for the *tk* CAAT mutations. However, on mutating the sixth residue from C to G (which creates a match to the *tk* sequence), a twofold reduction in expression efficiency was observed. In other words, both *tk* and MSV prefer a C residue immediately following the canonical 5′ CCAAT 3′ pentanucleotide.

The discovery of a region of dyad symmetry between the two Sp1-binding sites of the *tk* promoter provides a potential explanation for how Sp1 binding at the hexanucleotide of DSII might influence Sp1 binding to DSI. All prokaryotic repressor- and activator-binding sites that occur as dyad sequence signals are recognized by protein dimers that exhibit twofold rotational symmetry (Takeda et al. 1983). If the *tk* CAAT dyad is recognized by a rotationally symmetric protein, it would present the same image to both Sp1-binding sites. In theory, a protein bound in this position could bridge the two Sp1-binding sites, thereby facilitating an indirect interaction between them. Jones and Tjian (this volume) have identified such a protein in soluble extracts derived from HeLa cells. Peter Johnson and Barbara Graves have also identified a CAAT-binding protein from rat liver nuclei (unpubl.). Although neither the Tjian group nor our own group has established whether the CAAT-binding protein functions as a dimer, DNase I footprinting experiments show that the protein protects a region of DNA that covers both halves of the CAAT dyad (Jones and Tjian, this volume; P. Johnson and B. Graves, unpubl.).

If one envisions an interactive model whereby the CAAT-binding protein bridges the two Sp1-binding sites, one marked discrepancy remains. That is, the distance between the Sp1-binding site of DSII and the CAAT dyad (11 bp) is not equivalent to the distance between the Sp1-binding site of DSI and the CAAT dyad (21 bp). Since, however, the 10-bp difference in spacing represents roughly one full rotation of the DNA helix, it remains possible that interactions occur despite the spacing difference (by occurring preferentially on one face of the DNA helix).

This possibility was investigated in the following manner. A region of DNA between the CAAT dyad and the Sp1-binding site of DSI was deleted. In one case, 5 bp were removed, and in another case 10 bp were removed. Transcriptional assays of these two spacing-change mutants show that the 5-bp deletion reduces the expression efficiency of the *tk* promoter roughly 3-fold, whereas the 10-bp deletion increases expression efficiency roughly 4-fold (S. Eisenberg, R. Kingsbury, and S. McKnight, in prep.). Finally, when the 10-bp deletion mutant was established in the context of the G-to-C transversion mutant at residue −80 (which changes the dyad from 5′ ATTGGCGAAT 3′ to 5′ ATTGGCCAAT 3′), the resulting promoter is roughly 10-fold more efficient than normal.

The provocative conclusion that can be drawn from this study is that not only do proteins bind to activate the promoter of the HSV *tk* gene, but they further interact in specific manners to establish the precise "strength" of this promoter. This interpretation rests primarily on the observation that the "strength" of the *tk* promoter can be altered by more than an order of magnitude simply by changing the juxtaposition of existing protein-binding sites. Such a concept may bear relevance to the genetic apparatus of eukaryotic cells. It is possible that a fairly wide range of promoter "strengths" could be heritably built into different transcription units that utilize exactly the same set of transcriptional activators. It could further be predicted that a promoter could be induced in a temporal or tissue-specific manner by the controlled provision of a "bridge" that facilitates the interaction of a more common set of transcription factors.

REFERENCES

Dynan, W.S. and R.T. Tjian. 1983. The promoter-specific transcription factor Sp1 binds to upstream sequences in the SV40 early promoter. *Cell* **35:** 179.

Efstratiadis, A., J.W. Posakony, T. Maniatis, R.M. Lawn, C. O'Connell, R.A. Spritz, J.R. DeRiel, B.G. Forget, S.M. Weissman, J.L. Slightom, A.E. Blechl, O. Smithies, F.E. Baralle, C.C. Shoulders, and N.J. Proudfoot. 1980. The structure and evolution of the human β-globin gene family. *Cell* **21:** 653.

Johnson, A.D., A.R. Poteete, G. Lauer, R.T. Sauer, G.K. Ackers, and M. Ptashne. 1981. Repressor and *cro* − components of an efficient molecular switch. *Nature* **294:** 217.

McKnight, S.L. and R.C. Kingsbury. 1982. Transcriptional control signals of a eukaryotic protein-coding gene. *Science* **217:** 316.

McKnight, S.L., E.R. Gavis, R.C. Kingsbury, and R. Axel. 1981. Analysis of transcriptional regulatory signals of the HSV thymidine kinase gene: Identification of an upstream control region. *Cell* **25:** 385.

McKnight, S.L., R.C. Kingsbury, A. Spence, and M. Smith. 1983. The distal transcription signals of the herpesvirus tk gene share a common hexanucleotide control sequence. *Cell* **37:** 253.

Takeda, Y., D.H. Ohlendorf, W.F. Anderson, and B.W. Matthews. 1983. DNA-binding proteins. *Science* **221:** 1020.

Steroid Hormones Relieve Repression of the Ovalbumin Gene Promoter in Chick Oviduct Tubular Gland Cells

M.P. Gaub, A. Dierich, D. Astinotti, J.-P. LePennec, and P. Chambon

Laboratoire de Génétique Moléculaire des Eucaryotes du CNRS
Unité 184 de Biologie Moléculaire et de Génie Génétique de l'INSERM
Institut de Chimie Biologique, Faculté de Médecine
67085 Strasbourg, France

The chick oviduct is a particularly suitable system to study the molecular mechanisms regulating gene expression, because the synthesis of the major egg-white proteins (e.g., ovalbumin and conalbumin) can be modulated in the tubular gland cells by the administration and withdrawal of estrogens, progestins, glucocorticoids, and androgens, each acting through a distinct hormone receptor (see Shepherd et al. 1980 and references therein). After hormonal stimulation, there is an accumulation of the egg-white protein mRNAs, which results at least in part from an increased rate of transcription of the corresponding genes (see McKnight and Palmiter 1979 and references therein). The arrest of stimulation (withdrawal) leads to a cessation of synthesis of these mRNAs, which can be restimulated by the administration of any one of the four steroid hormones.

The molecular mechanisms by which steroid hormones induce transcription are poorly understood, but it is assumed that it is in some way mediated by the binding of nuclear steroid hormone–receptor complexes to regulatory sequences in the vicinity of the induced genes (e.g., see Chandler et al. 1983). However, it is not known whether the interaction between the hormone-receptor complex and the regulatory sequences leads to the activation of promoter elements otherwise intrinsically inactive (positive regulation) or to a relief of repression on promoter elements "constitutively" active unless repressed (negative regulation).

We have previously cloned the ovalbumin gene, which is specifically transcribed in the oviduct, and determined its structure (see

Heilig et al. 1982 and references therein). We have also shown that it is neither accurately nor efficiently expressed when transferred into a variety of nonoviduct cells (Breathnach et al. 1980; Chambon et al. 1984). Since there is no permanent cultured cell line derived from chicken oviduct or other chicken tissues that contains estrogen and progestin receptors and in which the cloned ovalbumin gene could be transferred to study its function and the mechanisms of its hormonal induction, we have microinjected recombinants containing the ovalbumin promoter region into the nucleus of primary cultured oviduct tubular gland cells.

RESULTS
A Microinjection Assay for Ovalbumin Promoter Activity
5′-flanking fragments of increasing lengths of the ovalbumin promoter (from position − 1 to positions − 132, − 295, − 425, or − 1348 upstream to the cap site; Fig. 1, the pTOT series) were coupled to the SV40 T-antigen coding sequence in such a manner that T-antigen expression is under the control of the ovalbumin promoter. pTOT series DNAs were microinjected into the nuclei of primary cultured oviduct tubular gland cells. Twenty-four hr later, the function of the ovalbumin promoter region was monitered by indirect immunofluorescence against T antigen. The validity of this promoter assay has been previously discussed (Moreau et al. 1981; Wasylyk et al. 1983). Since variations of fluorenscence intensity that cannot be validly measured are not taken into account, the assay

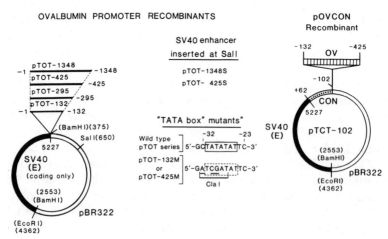

Figure 1

that scores the number of fluorescent nuclei rapidly reaches a saturation level as the amount of synthesized RNA increases. Large decreases in the efficiency of a promoter are accompanied by very significant decreases of the number of immunofluorescent nuclei, whereas increases of promoter efficiency above a certain level are reflected by only small increases in this number. The wild-type SV40 early-transcription-unit recombinant pSV1 (SV40 control in Tables 1 and 2), which functions in all cell types and is not sensitive to steroid hormones (our unpublished results), was microinjected in parallel to permit normalization of the data. The results of the pTOT series were expressed relative to the number of immunofluorescent cells obtained with pSV1 taken as 100% (usually 50–60% of the cells microinjected with pSV1 were T-antigen positive; see Tables 1 and 2, in which the average percentage of immunofluorescent tubular gland cells after microinjection of a given pTOT recombinant is underlined; the number of independent microinjection experiments and the extreme values for each recombinant are shown in parentheses and in brackets, respectively).

Effect of Steroid Hormone Antagonists on the Expression of the Ovalbumin Promoter Recombinants

All four pTOT recombinants were expressed to approximately the same levels after microinjection in nuclei of tubular gland cells maintained in a medium containing fetal calf serum (MFCS) (Table 1; data for pTOT-1348 are not shown). In contrast, none of the pTOT recombinants was expressed when microinjected into the fibroblasts that were also present in primary cultures of chicken oviduct cells, whereas the SV40 recombinant pSV1 was similarly expressed in both oviduct cell types (not shown). The possible effect of the steroid hormones present in FCS on expression of the ovalbumin promoter recombinant was investigated using the anti-estrogen, Tamoxifen, and a new compound, RU486, which is a specific anti-glucocorticoid in chicken (E. Baulieu and M. Govindan, pers. comm.). Tamoxifen and RU486 did not significantly affect the expression of pTOT-132 and pTOT-295, but the expression of pTOT-425 (and of pTOT-1348, not shown) was drastically inhibited by these antagonists. The residual level of activity may correspond to the presence of variable amounts of progestins in FCS since the expression of pTOT-425 was "reactivated" by the addition of progesterone. Microinjection of the TATA-box–mutated recombinants pTOT-132M and pTOT-425M, instead of their wild-type counterparts, resulted in an approximately fivefold decrease in expression,

Table 1
Effect of Anti-estrogens and Anti-glucocorticoids on the Activity of the Ovalbumin Promoter in Primary Cultures of Chicken Oviduct Cells Maintained in Medium with Fetal Calf Serum

Medium	control (SV40)	Recombinant		
		pTOT-132	pTOT-295	pTOT-425
MFCS[a]	100%[b]	66%[53–91](4)	53%[36–64](7)	53%[38–77](9)
MFCS + tamoxifen (10^{-6} M)	100%[c]	49%[46–54](3)	47%[33–64](4)	39%[23–54](4)
MFCS + RU486 (2×10^{-6} M)	100%[d]	47%[43–52](4)	41%[36–45](3)	26%[15–34](4)
+ [RU486 estradiol (10^{-8} M)]	100%	ND	ND	56%[51–60](2)
MFCS + [tamoxifen (10^{-6} M) RU486 (2×10^{-6} M)]	100%[e]	60%[44–84](6)	44%[36–49](4)	8%[0–16](8)
+ [tam+RU486 progesterone (10^{-9} M)]	100%[e]	ND	ND	50%[30–75](7)

[a] (MFCS) Medium with fetal calf serum. DNA: 100 copies/nucleus, 24 hr.
[b] 100% = 53% of injected cells became T antigen positive (8 exp.; extreme values: 43–59).
[c] 100% = 61% of injected cells became T antigen positive (3 exp.; extreme values: 54–67).
[d] 100% = 58% of injected cells became T antigen positive (4 exp.; extreme values: 54–65).
[e] 100% = 55% of injected cells became T antigen positive (6 exp.; extreme values: 46–71).
(ND) Not determined.

Table 2
Effect of Estradiol and Progesterone on the Activity of the Ovalbumin Promoter in Primary Cultures of Chicken Oviduct Cells Maintained in Steroid Hormone-stripped Medium

Recombinant	Culture medium					
	SM^a	SM + estradiol (10^{-8} M)^c	SM + progesterone (10^{-9} M)^d	SM + tamoxifen (10^{-6} M)^e	SM + RU486 (2×10^{-6} M)^f	SM + tamoxifen + RU486^g
Control (SV40)	100%^{b}	100%^{c}	100%^{d}	100%^{e}	100%^{f}	100%^{g}
pTOT-130	60%[42–100](7)	74%[58–84](4)	75%[60–85](5)	75%[58–91](2)	ND	60%[52–71](4)
pTOT-425	26%[10–42](13)	60%[34–91](8)	55%[34–75](7)	12%[8–17](4)	13%[5–20](4)	4%[0–8](7)
pTOT-295	60%[50–72](3)	64%[44–83](2)	65%[49–81](2)	ND	ND	53%[42–65](2)

a(SM) Steroid hormone-stripped medium. DNA: 100 copies/nucleus, 24 hr.
b100% = 50% of the injected cells became T antigen positive (15 exp.; extreme values: 43–60).
c100% = 49% of the injected cells became T antigen positive (9 exp.; extreme values: 36–74).
d100% = 53% of the injected cells became T antigen positive (9 exp.; extreme values: 34–68).
e100% = 50% of the injected cells became T antigen positive (3 exp.; extreme values: 45–55).
f100% = 55% of the injected cells became T antigen positive (4 exp.; extreme values: 45–67).
g100% = 48% of the injected cells became T antigen positive (8 exp.; extreme values: 36–68).
(ND) Not determined.

indicating that RNA synthesis was indeed initiated under the control of the ovalbumin promoter (Fig. 1, results not shown).

Expression of pTOT-425, but Not of pTOT-295 and pTOT-132, Is Blocked in the Absence of Steroid Hormones

To demonstate directly that expression of pTOT-425, but not of pTOT-295 and pTOT-132, is dependent on steroid hormones, primary cultured oviduct cells were maintained in a stripped medium (SM) from which these hormones were removed by treatment with dextran charcoal, which removes most, but not all, of the steroid hormones present in serum. Expression of pTOT-132 and pTOT-295 was insensitive to steroid hormone or antagonist additions (Table 2). On the contrary, microinjection of pTOT-425 resulted in a lower level of expression (26% in Table 2 instead of 53% in Table 1). Addition of both Tamoxifen and RU486 caused a further decrease of expression down to 4% on average, which most likely reflects a residual amount of progestins in the SM.

The repression exerted on expression of pTOT-425 in the absence of steroid hormones was relieved by the addition of estradiol or progesterone (Table 2). Both of them increased pTOT-425 expression to a level similar to that observed for pTOT-132 and pTOT-295. A similar level was obtained by adding progesterone to the SM in the presence of RU486 (not shown). The results obtained after addition of the specific glucocorticoid antagonist RU486 suggest that glucocorticoids can also relieve the repression exerted in the absence of steroid hormones. pTOT-1348 gave results identical to those obtained with pTOT-425 (not shown).

CONCLUSIONS

We have previously shown that none of the pTOT recombinants functions when microinjected into a variety of heterologous non-chicken or chicken cells, including the fibroblasts present in primary cultured oviduct or liver cells (Chambon et al. 1984 and unpublished results). However, any one of these recombinants was efficiently expressed in primary cultures of chicken embryo hepatocytes, irrespective of the presence of steroid hormones, although the ovalbumin gene is never expressed in chicken liver in spite of the presence of estradiol receptor. These results, together with those presented here, lead to the following conclusions (see also Fig. 2):

1. Chicken hepatocytes contain a cell-specific factor(s) that permits the constitutive activity of the microinjected ovalbumin promoter in the absence of steroid hormones. The corresponding

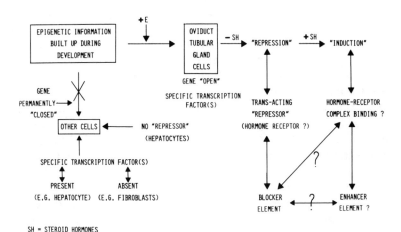

SH = STEROID HORMONES

Figure 2

control element(s) is contained within 132 bp upstream from the cap site.

2. Oviduct tubular gland cells contain a similar cell-specific factor(s), which permits the "constitutive" expression of the ovalbumin promoter region up to −295 bp upstream from the cap site, in the absence of steroid hormones. This cell-specific factor(s) is apparently not present in fibroblasts from primary cultures of liver or oviduct cells.

3. A negative control is exerted on the activity of the ovalbumin promoter region in oviduct tubular gland cells and involves sequences located between −295 and −425.

4. Hepatocytes lack the factor responsible for this negative control.

5. The presence of estradiol and/or progesterone (and most likely of glucocorticoids) relieves the "repression" in oviduct tubular gland cells.

Therefore the inactivity of the ovalbumin promoter in chicken tubular gland cells in the absence of steroid hormones appears to be due to the existence of a negative regulatory element (a blocker), involving sequences located between −295 and −425. Since there is no repression in hepatocytes, it is most likely that the blocker element is functional only in the presence of a *"trans*-acting" repressor. Thus the nonexpression of the ovalbumin gene in the liver of the whole animal is not due to the absence of a cell-specific transcription factor(s) that acts positively on promoter elements contained within 132 bp upstream from the cap site or to the presence of the repressor but must be due to a *cis*-acting negative mechanism that results from the "liver developmental history" of the gene and causes its permanent "closing." This closing mechanism can be bypassed by microinjection of plasmid recombinants. In other chicken cells, like fibroblasts, the cell-specific transcription factor(s) appears to be absent, since pTOT-132 is not expressed.

Our findings raise a number of interesting questions. First, what is the nature of the repressor and how do steroid hormones relieve its negative effect? Recent studies indicate that steroid hormone receptor molecules may be permanently located in the nucleus (see Schrader 1984 for references), which suggests that the unfilled receptor molecule may be involved in the repression. Second, do the various steroid hormone–receptor complexes interact with the blocker element or with a positive dominant regulatory element(s) when repression is relieved? Is such a positive regulatory element(s) an enhancer(s), as it has been shown to be in the regulation of MMTV promoter function by glucocorticoids (see Chambon et al. 1984 for references)? In this respect, it is interesting to note that inserting the − 132 to − 425 ovalbumin promoter region upstream from the + 62 to − 102 conalbumin promoter element (recombinant pOVCON in Fig. 1), which functions "constitutively" when microinjected alone (pTOT-102) into oviduct tubular gland cells (Chambon et al. 1984), results in "repression" of the latter in these cells in the absence of steroid hormones and in its induction in their presence. Thus it is clear that at least some of the sequences involved in the regulation of the ovalbumin promoter by steroid hormones are located between positions − 295 and − 425; this is upstream from the − 95 to − 222 region, which has been identified as responsible for a fivefold activation of the ovalbumin promoter by progesterone or estrogens (Dean et al. 1984). It is possible that the effect of this latter region has been missed in our present study because of the relative insensitivity of the immunofluorescence assay (see above). On the other hand, it is likely that the "blocker" effect of the − 295 to − 425 sequence has not been observed by Dean et al. (1984) because their ovalbumin-promoter chimeric recombinant contains the SV40 enhancer. We have indeed observed that the effect of the blocker sequence is not seen when the SV40 enhancer is introduced into pTOT-425 or pTOT-1348 (recombinants pTOT-425S and pTOT-1348S in Fig. 1; results not shown).

To our best knowledge, the present results constitute the first clear demonstration of the existence of a far-upstream negative regulatory element (a blocker) participating in the regulation of gene expression in a higher eukaryote. Together with enhancers (for review, see Chambon et al. 1984), blockers may provide the large number of combinatorial possibilities required for multifactorial regulation of gene expression in complex organisms.

ACKNOWLEDGMENTS

We thank Roussel-Uclaf (Paris) for providing RU486.

REFERENCES

Breathnach, R., N. Mantei, and P. Chambon. 1980. Correct splicing of a chicken ovalbumin gene transcript in mouse L cells. *Proc. Natl. Acad. Sci.* **77**: 740.

Chambon, P., A. Dierich, M.P. Gaub, S. Jakowlev, J. Jongstra, A. Krust, J.-P. LePennec, P. Oudet, and T. Reudelhuber. 1984. Promoter elements of genes coding for proteins and modulation of transcription by estrogens and progesterone. *Recent Prog. Horm. Res.* **40**: 1.

Chandler, V.L., B.A. Maler, and K.R. Yamamoto. 1983. DNA sequences bound specifically by glucocorticoid receptor in vitro render a heterologous promoter hormone responsive in vivo. *Cell* **33**: 489.

Dean, D.C., R. Gope, B.H. Knoll, M.E. Riser, and B.W. O'Malley. 1984. A similar 5'-flanking region is required for estrogen and progesterone induction of ovalbumin gene expression. *J. Biol. Chem.* **259**: 9967.

Heilig, R., R. Muraskowsky, and J.L. Mandel. 1982. The ovalbumin gene family. The 5' end region of the X and Y genes. *J. Mol. Biol.* **156**: 1.

McKnight, G.S. and R.D. Palmiter. 1979. Transcriptional regulation of the ovalbumin and conalbumin genes by steroid hormones in chick oviduct. *J. Biol. Chem.* **254**: 9050.

Moreau, P., R. Hen, B. Wasylyk, R. Everett, M.P. Gaub, and P. Chambon. 1981. The SV40 72 base pair repeat has a striking effect on gene expression both in SV40 and other chimeric recombinants. *Nucleic Acids Res.* **9**: 6339.

Schrader, W.T. 1984. New model for steroid hormone receptors? *Nature* **308**: 17.

Shepherd, J.C.W., E.R. Mulvihill, P.S. Thomas, and R.D. Palmiter. 1980. Commitment of chick oviduct tubular gland cells to produce ovalbumin mRNA during hormonal withdrawal and restimulation. *J. Cell Biol.* **87**: 142.

Wasylyk, B., C. Wasylyk, P. Augereau, and P. Chambon. 1983. The SV40 72 bp repeat preferentially potentiates transcription starting from proximal natural or subtitute promoter elements. *Cell* **32**: 503.

Additive Effects
of Two Distinct Activators
upon a Single Enhancer Element

D. DeFranco, R. Miesfeld, S. Rusconi,
and K.R. Yamamoto

Department of Biochemistry and Biophysics
University of California, San Francisco
San Francisco, California 94143

Complex, cell-specific patterns of gene expression are most likely generated by relatively few regulatory factors that operate according to combinatorial principles (Yamamoto 1985a). We have suggested that such multifactor control could be specified in part by simultaneous or ordered actions of multiple enhancers upon individual promoters (Yamamoto 1985b). Indeed, it appears that certain individual enhancer elements may themselves be activated by two or more separate *trans*-acting factors; for example, two segments from within the polyoma enhancer display differential activities in fibroblasts and teratocarcinoma cells (Herbomel et al. 1984). The nature of the putative *trans* activators responsible for these activities is unknown.

Transcriptional activation by glucocorticoid hormones is mediated by the sequence-specific binding of the glucocorticoid receptor protein to a class of enhancers termed glucocorticoid response elements (GREs). GREs were first detected within mouse mammary tumor virus DNA, where enhancer activity appears to be fully dependent upon the presence of bound receptor (Chandler et al. 1983; Zaret and Yamamoto 1984). We have discovered that the Moloney murine sarcoma virus (Mo-MSV) long terminal repeat (LTR) displays a GRE activity that is distinct from its previously described enhancer activity (Laimins et al. 1984). Exploiting the known properties of the glucocorticoid receptor, we have tested the possibility that the Mo-MSV enhancer is a complex element that can exhibit two activities distinguishable even within a single cell type. We have also cloned a portion of the receptor cDNA and genomic sequences, in order eventually to understand in detail the structure and function of this enhancer-activating protein.

Mapping Two Mo-MSV Enhancer Activities

Khoury and co-workers (Laimins et al. 1984) have described an enhancer element within a 73–72-bp tandem duplication beginning 183 bp upstream from the Mo-MSV transcription initiation site within the LTR (Fig. 1). This activity, which we term S_a, appears preferentially operative in murine cells, and a putative activating/regulatory factor has been inferred from certain transfection studies (Schöler and Gruss 1984). Consistent with the mapping data, Table 1 shows that the inact Mo-MSV LTR fused to the herpes simplex virus (HSV) thymidine kinase coding sequences (*tk*) yields 25-fold more properly initiated *tk* mRNA than a fusion lacking Mo-MSV sequences upstream of – 110 bp from the Mo-MSV transcription initiation site. Treatment of the transfected cultures with 0.1 μM dexamethasone produced an additional 4-fold increase in activity from the fusions with the intact LTR, but not from the 5′ – 110 deletion plasmid. This suggests that a GRE activity, here termed S_g, is also contained within the Mo-MSV LTR upstream from – 110 bp.

A 298-bp fragment bearing Mo-MSV LTR sequences extending upstream of – 146, which contains full S_a activity, was cloned in both orientations either upstream or downstream of the intact *tk* promoter in a chloramphenicol acetyltransferase (CAT) expression plasmid. Table 1 shows that in every case both S_a and S_g activities are observed after transfection into XC rat fibroblast cells, indicating that S_a and S_g coreside within the 298-bp fragment. In contrast, S_a activity was undetectable when the same plasmids were transfected into AR4-2J rat pancreas cells, whereas S_g was strongly activated by dexamethasone in both cell lines. Taken together, these results indicate that S_a and S_g are distinguishable and independent enhancer activities, both mapping to the 73–72-bp tandem repeat region.

Nuclease footprinting experiments with purified glucocorticoid receptor revealed three discrete footprints of equivalent strength

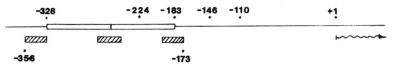

Figure 1 Diagram of the Mo-MSV LTR, showing the tandem 73–72-bp repeats associated with S_a enhancer activity (open boxes) and the positions of three glucocorticoid receptor footprints (hatched boxes). Transcription initiates at + 1 and proceeds rightward.

Table 1
Identification of Two Distinguishable Activities within the Mo-MSV Enhancer

Experiment	Cell line	Relative position and orientation of enhancer	Promoter	− enhancer	Relative transcriptional activity		
					enhancer S_a	enhancers $S_a + S_g$	$(S_a + S_g)/S_a$
1	XC fibroblast	5'→	Mo-MSV	1	25	100	4
2	XC fibroblast	5'→	tk	1	22	171	7.8
		5'←	tk	1	4.7	36	7.7
		3'→	tk	1	29.2	153	5.2
		3'←	tk	1	23	114	4.9
3	XC fibroblast	5'→	SV40	1	8.9	41.3	4.7
		5'Δ224→	SV40	1	0.35	8.2	23
4	AR4-2J pancreas	5'→	tk	1	1	65	65
		5'←	tk	1	1	7.8	7.8
		3'→	tk	1	1	16	16
		3'←	tk	1	1	39.4	39.4

Results of four separate transient transfection experiments are shown; relative transcription activities were inferred from assays of CAT activity in experiments 2, 3, and 4 and from primer extension analysis of properly initiated RNA in experiment 1. The enhancer orientation was either the same as (rightward arrow) or opposite to (leftward arrow) that seen with its homologous promoter. In each case, a fragment bearing the intact Mo-MSV enhancer was used, except with 5'Δ224 (experiment 3), which lacks Mo-MSV LTR sequences upstream of −224 (constructed and provided by P. Gruss). S_a and $S_a + S_g$ denote enhancer activities in cells transfected in the absence and presence of 0.1 μM dexamethasone, respectively; the magnitude of the hormone effect (i.e., S_g activity alone) is represented by the ratio $S_a + S_g/S_a$. Dexamethasone had no effect on promoter activities of constructions lacking the enhancer.

134

within the fragment containing S_a and S_g activities. Each footprint contains a consensus octanucleotide that has been implicated in specific receptor-binding and GRE enhancer functions (Payvar et al. 1983; DeFranco et al. 1985); the positions of the three footprints imply that S_a and S_g activities may be interdigitated (Fig. 1). Consistent with this view, plasmids lacking sequences upstream of -224 display no S_a activity but retain partial S_g activity, presumably mediated by the single receptor-footprint sequence not deleted in that mutant (Fig. 1; Table 1).

Receptor Coding Sequences

Rat glucocorticoid receptor cDNA fragments were initially recovered by polysome immunoenrichment, followed by cDNA library construction and screening by differential hybridization (Miesfeld et al. 1984). Overlapping clones that together contain the distal 5.2 kb of an ∼7-kb receptor mRNA detected in rat, mouse, and human cells have been isolated. DNA sequence analysis and expression of fusion products in *Escherichia coli* reveal that this fragment contains a 1.4-kb open reading frame, which could account for about half of the amino acid coding sequence of the 94-kD receptor, followed by 3.8 kb of 3′ untranslated sequence. Using hybridization probes from the coding sequence, we have shown that the receptor is a highly conserved, unique gene and that it resides on mouse chromosome 18. Preliminary data suggest that the carboxyl half of the receptor protein contains the DNA- and steroid-binding domains. Consistent with genetic and biochemical studies of the receptor protein, functional regions appear to be discrete and linearly arranged along the primary sequence. Attempts are now underway to recover DNA fragments encoding the amino half of the receptor; conceivably, this region may contain an activity uniquely involved in the mechanism of enhancement.

DISCUSSION

Our detection of a glucocorticoid-dependent enhancer activity within the Mo-MSV LTR might explain the finding that dexamethasone increases the intracellular accumulation of Mo-MSV RNA in an infected mink cell line (Lowy and Scolnick 1978). In any case, the capacities to regulate S_g activity and to determine directly the sites of receptor : DNA interaction have provided strong evidence that the S_a and S_g activities within the Mo-MSV enhancer are distinguishable, independent, and separable. These conclusions are supported by assessment of the two activities in a deletion mutant lacking much of the 73–72-bp duplication and in transfections of different cell types.

Taken together, the results imply that the S_a and S_g activities are approximately additive. Thus, assuming that most of the Mo-MSV enhancers resident in a given cell become activated in response to the appropriate signals, the glucocorticoid receptor and the putative factor that mediates S_a activity can apparently bind simultaneously to the enhancer region, presumably in an interdigitated array suggested by the receptor footprints. Although the properties of the S_a factor are unknown, the glucocorticoid receptor is a 94-kD protein that appears to bind to DNA as a multimeric complex in vitro (Payvar et al. 1983) and to create discrete alterations in chromatin structure upon binding in vivo (Zaret and Yamamoto 1984). A detailed understanding of receptor structure and its activation of S_g, together with studies on the nature of the S_a factor and the potential formation of a heteromeric enhancer complex, may reveal some important mechanistic features of transcriptional enhancement and its role in multifactor gene regulation.

REFERENCES

Chandler, V.L., B.A. Maler, and K.R. Yamamoto. 1983. DNA sequences bound specifically by glucocorticoid receptor in vitro render a heterologous promoter hormone responsive in vivo. *Cell* 33: 489.

DeFranco, D., Ö. Wrange, J. Merryweather, and K.R. Yamamoto. 1985. Biological activity of a glucocorticoid-regulated enhancer: DNA sequence requirements and interactions with other transcriptional enhancers. *UCLA Symp. Mol. Cell. Biol.* 20: 305.

Herbomel, P., B. Bourachot, and M. Yaniv. 1984. Two distinct enhancers with different cell specificities coexist in the regulatory region of polyoma. *Cell* 39: 653.

Laimins, L.A., P. Gruss, R. Pozzatti, and G. Khoury. 1984. Characterization of enhancer elements in the long terminal repeat of Moloney murine sarcoma virus. *J. Virol.* 49: 183.

Lowy, D.R. and E.M. Scolnick. 1978. Glucocorticoids induce focus formation and increase sarcoma viral expression in a mink cell line that contains a murine sarcoma viral genome. *J. Virol.* 25: 157.

Miesfeld, R., S. Okret, A.-C. Wikström, Ö. Wrange, J.-Å. Gustafsson, and K.R. Yamamoto. 1984. Characterization of a steroid receptor gene and transcripts in wild and mutant cells. *Nature* 312: 779.

Payvar, F., D. DeFranco, G.L. Firestone, B. Edgar, Ö. Wrange, S. Okret, J.-Å. Gustafsson, and K.R. Yamamoto. 1983. Sequence-specific binding of glucocorticoid receptor to MTV DNA at sites within and upstream of the transcribed region. *Cell* 35: 381.

Schöler, H.R. and P. Gruss. 1984. Specific interaction between enhancer-containing molecules and cellular components. *Cell* 36: 403.

Yamamoto, K.R. 1985a. Steroid receptor–regulated transcription of specific genes and gene networks. *Annu. Rev. Genet.* (in press).

———. 1985b. Hormone-dependent transcriptional enhancement and its implications for mechanisms of multifactor gene regulation. *Symp. Soc. Dev. Biol.* **43:** (in press).

Zaret, K.S. and K.R. Yamamoto. 1984. Reversible and persistent changes in chromatin structure accompany activation of a glucocorticoid-dependent enhancer element. *Cell* **38:** 29.

Cell-specific Expression of Pancreas Genes: *cis*-acting Control Elements Located within 5'-flanking DNA

T. Edlund,* † M.D. Walker, † A.M. Boulet, †
C. Erwin, † and W.J. Rutter †

*Department of Applied Cell and Molecular Biology
University of Umea, Umea S901 87, Sweden

†Hormone Research Institute, University of California, San Francisco
San Francisco, California 94143

The patterns of differential gene expression observed in various differentiated cells of eukaryotic organisms are attributable in large measure to control at the level of initiation of RNA synthesis (Darnell 1982). Promoters (McKnight and Kingsbury 1982) and enhancers (Khoury and Gruss 1983), two classes of DNA elements with somewhat different properties, apparently cooperate in the control of initiation of transcription. Promoter elements are located in the immediate vicinity of the transcription start (cap) site (typically within 110 bp) and are sensitive to orientation and precise position with respect to the cap site. Enhancer elements, which were first discovered in DNA tumor viruses and subsequently in other viral and cellular genes, have the distinguishing feature of being much less influenced by orientation and position. The properties of enhancers and promoters are sufficiently distinct so as to suggest a fundamental difference in their mechanisms of action.

Our laboratory has been interested in elucidating the mechanisms involved in cell-specific gene expression using the mammalian pancreas as a model system. To this end we have isolated and sequenced 13 genes that are selectively expressed in the pancreas. These include the genes for amylase (Crerar et al. 1983), chymotrypsin (Bell et al. 1984), trypsin (Craik et al. 1984) and elastase (Swift et al. 1984), products of the exocrine pancreas, and insulin (Bell et al. 1980) and somatostatin (Shen and Rutter 1984), products of the endocrine pancreas. We have established that DNA sequences located within the proximal 300 bp of 5'-flanking DNA play an important role in directing the cell-specific expression of insulin and chymotrypsin genes in transfected cell lines derived from endocrine and exocrine pancreas, respectively (Walker et al. 1983). In this report we document the presence of cell-specific transcriptional enhancers within the 5'-flanking DNA of several pancreatic genes and determine the location of these elements by 5' and 3' deletion analysis. Furthermore, we present evidence that DNA sequences within the promoter region of the insulin gene, although devoid of enhancer activity, can contribute to cell-specific expression.

RESULTS AND DISCUSSION

To conveniently identify and characterize *cis*-acting transcriptional control elements, we constructed the plasmid pTE1 (T. Edlund et al., in prep.), which contains the herpes simplex virus (HSV) thymidine kinase (*tk*) promoter upstream from the chloramphenicol acetyltransferase gene, *cat* (Gorman et al. 1982). This plasmid simplifies the assessment of both position effects and the construction of sets of 5'- and 3'-deletion mutants of enhancer fragments. When a 5'-flanking DNA fragment of the rat insulin I gene (−410 to +51) was inserted into pTE1 immediately upstream from the *tk* promoter but in opposite orientation, there was a 25-fold enhancement of *tk* promoter activity in HIT cells (hamster insulin-producing cell line) but no effect in BHK cells (hamster fibroblasts). This fragment was also active (albeit at lower levels) when placed either 600 bp further upstream (3-fold) or downstream from *cat* (11-fold). A similar pattern was seen when 5'-flanking DNA sequences from the chymotrypsin gene (−275 to −93) and amylase gene (−234 to −41) were introduced to pTE1. The chymotrypsin fragment had effects of 48-fold (5' proximal), 2-fold (5' distal), and 5-fold (3') when tested in AR4-2J cells (rat exocrine pancreas tumor line) but little or no effect in XC cells (rat fibroblasts). The amylase fragment showed effects of 33-fold (5' proximal) and 10-fold (5' distal) in AR4-2J cells but no effect in XC cells.

The location of important sequences within the flanking DNA was determined by deletion analysis. Exonuclease III was used to obtain sets of 5' and 3' deletions of the three pancreatic enhancers (Table 1). In each case the analysis suggested a minimal sequence that by itself should act as a cell-specific enhancer. In the case of the insulin-flanking DNA, the −103 to −249 fragment was predicted to have enhancer activity. Indeed, when such a fragment was inserted into pTE1 immediately upstream from the *tk* promoter, it increased expression by approximately 20-fold in an orientation-independent manner. Similarly for the amylase gene, deletion anal-

Table 1
Deletion Analysis of Pancreatic Enhancers

Insulin			Chymotrypsin				Amylase		
5'	3'	activity	5'	3'	activity		5'	3'	activity
−410	+51	100	−711	−3	100		−234	−41	100
−333	+51	147	−275	−3	97		−199	−41	154
−287	+51	81	−225	−3	102		−175	−41	123
−249	+51	79	−192	−3	9		−154	−41	139
−219	+51	22	−184	−3	7		−140	−41	5
−113	+51	4							
−410	−103	150	−275	−93	100		−234	−41	94
−410	−150	37	−275	−109	145		−234	−77	30
−410	−198	38	−275	−113	61		−234	−115	38
−410	−249	3	−275	−150	70		−234	−159	4
			−275	−170	17		−234	−179	3
tk_p only		2.5	tk_p only		2-3		tk_p only		2-3

5'- and 3'-deletion analysis of enhancer activity of rat insulin I gene, rat chymotrypsin B gene, and rat pancreatic amylase gene. DNA fragments from these genes were inserted into the plasmid pTE1. Deletion mutants were generated by treatment with exonuclease III. The activity of 5' deletions from the insulin and amylase gene fragments was determined with fragments inverted with respect to the tk promoter (tk_p). The series of 5' deletions of the chymotrypsin gene was generated with the chymotrypsin promoter rather than the tk promoter. For 3' deletions, fragments were oriented in the same way as the tk promoter. Activity was determined in the appropriate cell type, i.e., HIT cells (insulin producing) for the insulin enhancer and AR4-2J cells (amylase and chymotrypsin producing) for amylase and chymotrypsin genes.

ysis predicted that a fragment from −115 to −154 should display enhancer activity. A slightly larger fragment was synthesized (−109 to −163) and indeed was found to possess cell-specific enhancer activity.

Comparison of the pancreas enhancers with viral enhancers reveals no striking sequence similarities. An enhancer-core consensus sequence first identified by Weiher et al. (1983) is not present in the 5'-flanking sequence of the amylase or chymotrypsin genes. A related sequence is present, centered at nucleotide −312 in the rat insulin I gene, but can be removed with no dramatic loss of enhancer activity. On the other hand, comparisons among the 5'-flanking DNA sequences of exocrine pancreas genes show some distinctive homologies. Significantly, these homologies are found in regions of the chymotrypsin and amylase genes, which are shown by our analysis to be essential for cell-specific enhancer activity (A.M. Boulet et al., in prep.). In the other exocrine pancreas genes, these sequences (the exocrine pancreas enhancer consensus sequences) are located at different precise positions but in the same general region of the 5'-flanking DNA (−90 to −250).

The detection of insulin enhancer activity upstream of −103 does not eliminate the possibility that sequences downstream from this position play an important role in cell-specific expression. Indeed, we have previously observed a striking sequence conservation not only within the enhancer region but also in the promoter region of different mammalian insulin genes (Bell et al. 1982). To test for a cell-specific function for sequences within the promoter region, we carried out an "enhancer replacement" experiment by substituting the insulin enhancer with an enhancer derived from Moloney murine sarcoma virus (Mo-MSV). This enhancer is active in both HIT and BHK cells. The activity of recombinants containing the Mo-MSV enhancer linked to the insulin promoter region was measured in HIT cells and BHK cells and the relative expression was compared with several control constructs (Table 2). A clear (∼8-fold) cell-specific preference was observed. This is not attributable to an intrinsic enhancer activity of the insulin promoter region (see Table 1), nor is it due to cooperation between the Mo-MSV enhancer and a residual cryptic enhancer activity of the insulin promoter region (data not shown). More recently, we have precisely deleted all transcribed sequences (+1 to +51) from such enhancer replacement constructs and find no significant effect on cell-specific preference, suggesting that the activity is mediated at the level of initiation of transcription.

Our results show that 5'-flanking regions can direct cell-specific

Table 2
Cell Specificity Associated with the Insulin Gene Promoter Region

| Construct | Distal 5′ | Proximal 5′ | Relative activity | | Ratio |
			HIT	BHK	
1	MSV_E	TK_p	1.00	1.00	1.0
2	MSV_E	MSV_p	3.80	5.21	0.7
3	RSV_E	RSV_p	1.65	2.33	0.7
4	–	β-actin	1.08	2.73	0.4
5	MSV_E	Ins_p	0.41	0.05	8.2

The ability of an enhancer replacement plasmid (no. 5) containing the Mo-MSV enhancer immediately upstream from the insulin promoter region (−113 to +51) to direct the expression of a contiguous CAT gene was measured in HIT and BHK cells. Activities were compared with those obtained with control constructs as follows: (1) Mo-MSV enhancer upstream from *tk* promoter; (2) Mo-MSV enhancer upstream from Mo-MSV promoter; (3) RSV enhancer upstream from RSV promoter; (4) rat β-actin gene promoter.

expression of pancreas genes. Included within these regions are enhancer elements that are selectively active in the appropriate cell type. In addition, we present evidence that sequences within the promoter region may also contribute to cell-specific expression. We postulate that the activity of these *cis*-acting elements is controlled by the presence of *trans*-acting transcriptional regulatory factors, which we term differentiators, to indicate their direct role in generating cellular phenotype. The sequence homologies observed in the 5′-flanking regions of pancreatic exocrine genes suggests that the pancreatic exocrine phenotype may be controlled by a small family of differentiators.

ACKNOWLEDGMENTS
We thank the following for gifts of plasmids: Drs. G. An, C. Gorman, O. Karlsson, P. Luciw, U. Nudel, R. Tjian, and K. Zaret. We also thank Dr. C. Craik and J. Barnett for oligonucleotides, J. Turner, C. Griffin, and A. Backman for technical assistance, and L. Spector for preparation of the manuscript. We acknowledge the constructive suggestions of our colleagues in the Rutter laboratory. This work was supported by National Institutes of Health grant AM-21344 and GM-28520 and a grant from the March of Dimes.

REFERENCES
Bell, G.I., M.J. Selby, and W.J. Rutter. 1982. The highly polymorphic region near the human insulin gene is composed of simple tandemly repeating sequences. *Nature* **295:** 31.

Bell, G.I., R.L. Pictet, W.J. Rutter, B. Cordell, E. Tischer, and H.M. Goodman. 1980. Sequence of the human insulin gene. *Nature* **284:** 26.

Bell, G.I., C. Quinto, M. Quiroga, P. Valenzuela, C.S. Craik, and W.J. Rutter. 1984. Isolation and sequence of a rat chymotrypsin B gene. *J. Biol. Chem.* **259:** 14265.

Craik, C.S., Q.L. Choo, G.H. Swift, C. Quinto, R.J. MacDonald, and W.J. Rutter. 1984. Structure of two related rat pancreatic trypsin genes. *J. Biol. Chem.* **259:** 14255.

Crerar, M.M., W.F. Swain, R.L. Pictet, W. Nikovits, and W.J. Rutter. 1983. Isolation and characterization of a rat amylase gene family. *J. iol. Chem.* **258:** 1311.

Darnell, J.E. 1982. Variety in the level of gene control in eukaryotic cells. *Nature* **297:** 365.

Gorman, C., L.F. Moffat, and B.H. Howard. 1982. Recombinant genomes which express chloramphenicol acetyltransferase in mammalian cells. *Mol. Cell. Biol.* **2:** 1044.

Khoury, G. and P. Gruss. 1983. Enhancer elements. *Cell* **33:** 313.

McKnight, S.L. and R. Kingsbury. 1982. Transcriptional control signals of a eukaryotic protein-coding gene. *Science* **217:** 316.

Shen, L.P. and W.J. Rutter. 1984. Sequence of the human somatostatin I gene. *Science* **224:** 168.

Swift, G.H., C.S. Craik, S.J. Stary, C. Quinto, R.G. Lahaie, W.J. Rutter, and R.J. MacDonald. 1984. Structure of the two related elastase genes expressed in the rat pancreas. *J. Biol. Chem.* **259:** 14271.

Walker, M.D., T. Edlund, A.M. Boulet, and W.J. Rutter. 1983. Cell-specific expression controlled by the 5'-flanking region of insulin and chymotrypsin genes. *Nature* **306:** 557.

Weiher, H., M. Konig, and P. Gruss. 1983. Multiple point mutations affecting the simian virus 40 enhancer. *Science* **219:** 626.

Metal Regulatory Elements in Metallothionein Gene Promoters

R.D. Palmiter, G.W. Stuart, and P.F. Searle

Howard Hughes Medical Institute, Department of Biochemistry
University of Washington, Seattle, Washington 98195

Metallothionein (MT) genes are transcriptionally regulated by a variety of metals, including Zn, Cd, Cu, Ag, Hg, Co, Ni, and Bi (Durnam and Palmiter 1981, 1984). In mice there are two closely linked MT genes, designated MT-I and MT-II (Searle et al. 1984), whereas in primates the MT gene family is composed of several members (Karin and Richards 1982). The two mouse MT genes are regulated in a strictly coordinate manner in response to Zn, Cu, and Cd, suggesting that the two genes may be regulated by the same metal-binding proteins (Yagle and Palmiter 1985).

To identify regions of the MT gene that are responsible for metal regulation, we initially fused the MT promoter and 5'-flanking region to the thymidine kinase structural gene (*tk*) of herpes simplex virus. We showed that viral TK enzyme activity was inducible by metals after transfer of this construct into tissue-culture cells, mouse eggs, or mice (Brinster et al. 1982; Mayo et al. 1982; Palmiter et al. 1982). Analysis of a series of 5' deletions revealed that all of the elements necessary for maximal induction by metals lie within 217 bp of the transcription start site. Further deletions produced a gradual decline in both basal and metal-inducible expression until only a very low, uninducible level of expression was reached when 50 bp of 5' sequence remained.

A series of 3' deletion, internal deletion, and linker-scanning mutants was generated and tested by either microinjection into mouse eggs or transfection into baby hamster kidney (BHK) cells. Although many of the mutants partially decreased basal and/or induced expression, there were no sharp transitions from an inducible to an uninducible phenotype (Stuart et al. 1984; Searle et al. 1985). The results suggested that two or more metal regulatory elements (MREs) lie within the MT promoter region. By similar methods, Carter et al. (1984) and Karin et al. (1984) concluded that there are at least two MREs in the mouse MT-I and human MT-IIA promoters, respectively. By comparing the sequences of several mammalian MT promoters to those of the mutants that had the most dramatic effects on inducible expression, we identified a conserved sequence that is centered at about −50 and several related sequences upstream that are present in all mammalian MT gene promoters (Fig. 1A). In the mouse MT-I promoter, five potential MRE sequences are identified. They are designated MRE-a through MRE-e; their sequences and locations are shown in Figure 1, C and E.

To ascertain whether these sequences can function independently as MREs, we synthesized a pair of complementary oligonucleotides (Fig. 1B) that generate a DNA sequence including the MRE and inserted them into the promoter of the *tk* gene. Only 13 of the 17 nucleotides within the synthetic MRE-a′ that we tested match the MT-I promoter sequence (Fig. 1B); the other 4 nucleotides allow easy cloning into *Bam*HI and *Bgl*II restriction sites, regenerating the site at one end of the insert. We tested the effect of single MRE-a′ elements inserted into several different positions within the *tk* promoter for their ability to confer metal regulation. Using the mouse egg assay system, a single MRE was capable of conferring metal regulation to this heterologous promoter, whereas with the BHK transfection assay, single elements gave a negligible response. However, in both assay systems, two or more MREs consistently conferred metal regulation to the *tk* promoter (Stuart et al. 1984; Searle et al. 1985). The differences in results obtained with the two assay systems are consistent with those obtained when assaying 5′ deletions that retain only one MRE and probably reflect the relative amounts of template, transcription factors, and regulatory proteins in the two systems.

We then examined whether the relative positions or orientations of the multiple MREs were important for metal inducibility in transfected BHK cells. We tested pairs of MREs inserted into various positions within the *tk* promoter, either in tandem or in inverted orientation, or separated from each other by up to 180 bp. The general conclusion is that all combinations of two MRE-a′ sequences effectively confer metal regulation to the *tk* promoter if they are located within about 100 bp of each other, including a construction in which the closest MRE was 62 bp from the TATA box. The induction was generally 2-fold to 3-fold for two MREs, although the absolute level of expression depended upon whether the *tk* promoter elements were disrupted by the insertions. By inserting three or four MRE-a′ elements, it was possible to achieve up to a 10-fold induction (Searle et al. 1985).

We also tested MRE-a, which extends the homology to the MT-I sequence by 4 nucleotides in the 3′ direction (Fig. 1D), but this se-

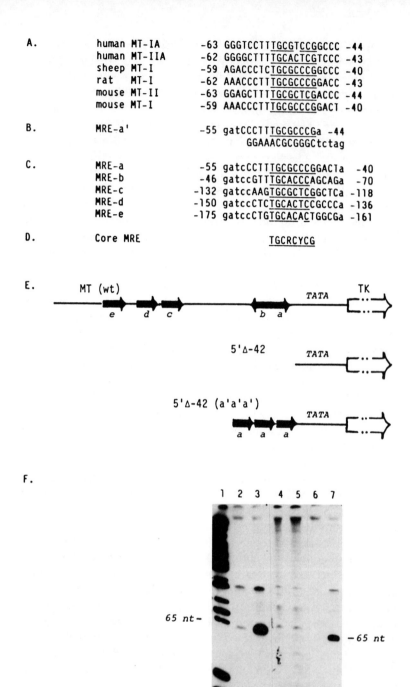

Figure 1 *See facing page for legend.*

quence was only marginally better than MRE-a'. Thus, we conclude that a functional MRE is contained within the 13-base sequence CCCTTTGCGCCCG. From mutational analysis and a homology search, Karin et al. (1984) suggested that the sequence ACTCGT-CCCGGCTC is the proximal MRE in the human MT-IIA gene. However, when we tested two copies of this sequence in our assay, they did not allow metal induction. We believe that their proposed regulatory sequence deletes essential nucleotides from the 5' end (see Fig. 1A for our alignment of the human and mouse sequences).

In another series of BHK transfection experiments, we inserted one to four copies of MRE-a' just upstream of the *tk* or MT TATA-box regions and deleted all other known 5' promoter elements (Fig. 1E). In both cases, the additional MRE-a' elements had little or no effect upon basal activity (which was very low) but promoted a progressive increase in expression in response to Zn as more elements were inserted upstream of either TATA box. Figure 1F shows the results of an S1 nuclease protection assay that was used to map the 5' end of the mRNA produced from one of these constructs. The single-stranded probe was made from 5'Δ-42 plasmid, end-labeled at the *Bgl*II site (+64). The wild-type MT-I promoter showed a band corresponding to correct start site that is strongly inducible by Zn (lanes 2 and 3). (The upper band corresponds to the point of divergence between the probe and the plasmid and thus represents the

Figure 1 (*see facing page*)   The MREs of MT gene promoters. (*A*) The homology of the potential MRE sequences centered at about −50 of various mammalian MT genes; nucleotides identical to the core MRE (see *D*) are underlined. (*B*) The pair of synthetic oligonucleotides that constitute MRE-a'; lower-case letters indicate deviations from the mouse MT sequence; they are included to regenerate convenient restriction sites when inserted into *Bam*HI or *Bgl*II sites. (*C*) The sequences of synthetic MREs a–e that were tested for metal inducibility. Lower-case lettering and underlining are defined above; the locations of the sequences relative to the transcription start site are indicated. (*D*) The core MRE sequence deduced by homology search, mutation analysis, and testing of the various synthetic MREs. (*E*) Diagram of the mouse MT-I gene promoter region, showing locations of the various MREs (lettered arrows). A truncated MT promoter with a *Bam*HI site at −42 and the same construct with three synthetic MRE-a' sequences inserted upstream are also shown. (*F*) S1 nuclease protection assay of RNA transcripts produced in BHK cells transfected with *tk* genes containing the MT promoters shown in *E*: (Lane *1*) *Hpa*II markers; (lanes *2* and *3*) wild-type MT promoter minus and plus Zn, respectively; (lanes *4* and *5*) truncated MT promoter minus and plus Zn, resectively; (lanes *6* and *7*) truncated MT promoter with three MREs minus and plus Zn, respectively. The major protected band (65 nt) corresponds to the size of the RNA expected if transcription were to start at the normal cap site.

sum of all upstream starts.) A low amount of random initiation occurred within the promoter region of 5'Δ-42, and these transcripts were not inducible by Zn (lanes 4 and 5). However, when three MREs were inserted, random initiation events were suppressed and correctly initiated transcripts were strongly inducible by Zn (lanes 6 and 7). These experiments are most easily interpreted as indicating that a metal-binding protein binds to MRE-a sequences to enhance transcription, i.e., the regulatory protein is a positively acting transcription factor.

We have begun to ask whether the other MRE-like sequences that are located within the mouse MT-I promoter region are also active by inserting pairs of these elements into the *tk* promoter and testing them by transfection into BHK cells. The results indicate that MRE-a, MRE-b, and MRE-c stimulate transcription in the presence of Zn, whereas MRE-d and MRE-e have marginal activity. Interestingly, insertion of MRE-d results in a high basal activity that is not very inducible. The MRE-d element that we tested includes a TCCGCCCA sequence, which matches an Sp1 binding site and may be responsible for the basal activity (K. Jones and R. Tjian, unpubl.). However, the A in this Sp1 sequence marks the divergence of the oligomer from the wild-type MRE-d sequence. Insertion of a G before the A extends the homology to MT-I by two nucleotides, eliminates the basal activity, and allows metal-inducible expression (G.W. Stuart et al., in prep.).

From the conservation of particular nucleotides in the MREs, a comparison of the MRE sequences that are most effective in this assay, and the effects of mutants that impinge upon MREs, we have refined our ideas about what nucleotides within the consensus sequence are most likely to be involved in binding metal regulatory protein(s). The working hypothesis is that the sequence TGCRCNCG provides the core of a functional MRE. We are now in the process of testing variants of the MRE sequence to further define the nucleotides required for function.

DISCUSSION

All natural MT gene promoters contain several copies of the core MRE sequence described above. Deletion analysis has indicated that there are at least two functional MREs in both the mouse MT-I and human MT-IIA promoters. Insertion of synthetic MRE-like sequences into a heterologous promoter has revealed at least four functional MREs within the mouse MT-I promoter. It is likely that a metal-binding protein interacts with these sequences to induce transcription. This metal-binding protein is thought to be a positive regulator of transcription rather than a repressor because (1) no mutants have been found that increase basal expression, (2) competition experiments with excess regulatory region depress the induction rather than increase basal expression (Seguin et al. 1984), and (3) insertion of MREs upstream of a TATA box is almost neutral in the absence of metals and greatly enhances transcription in the presence of metals. Definitive proof must await purification of the protein(s) involved in metal regulation of the MT promoters.

REFERENCES

Brinster, R.L., H.Y. Chen, R. Warren, A. Sarthy, and R.D. Palmiter. 1982. Regulation of metallothionein–thymidine kinase fusion plasmids injected into mouse eggs. *Nature* **296**: 39.

Carter, A.D., B.K. Felber, M.-J. Walling, M.-F. Jubier, C.J. Schmidt, and D.H. Hamer. 1984. Duplicated heavy metal control sequences of the mouse metallothionein-I gene. *Proc. Natl. Acad. Sci.* **81**: 7392.

Durnam, D.M. and R.D. Palmiter. 1981. Transcriptional regulation of the mouse metallothionein-I gene by heavy metals. *J. Biol. Chem.* **256**: 5712.

———. 1984. Induction of metallothionein-I mRNA in cultured cells by heavy metals and iodoacetate: Evidence for gratuitous inducers. *Mol. Cell. Biol.* **4**: 484.

Karin, M. and R.I. Richards. 1982. Human metallothionein genes – primary structure of the metallothionein-II gene and a related processed gene. *Nature* **299**: 797.

Karin, M., A. Haslinger, H. Holtgreve, R.I. Richards, P. Krauter, H.M. Westphal, and M. Beato. 1984. Characterization of DNA sequences through which cadmium and glucocorticoid hormones induce human metallothionein-IIA gene. *Nature* **308**: 513.

Mayo, K.E., R. Warren, and R.D. Palmiter. 1982. The mouse metallothionein-I gene is transcriptionally regulated by cadmium following transfection into human or mouse cells. *Cell* **29**: 99.

Palmiter, R.D., H.Y. Chen, and R.L. Brinster. 1982. Differential regulation of metallothionein–thymidine kinase fusion genes in transgenic mice and their offspring. *Cell* **29**: 701.

Searle, P.F., G.W. Stuart, and R.D. Palmiter. 1985. Building a metal responsive promoter with synthetic regulatory elements. *Mol. Cell. Biol.* **5**: 1480.

Searle, P.F., B.L. Davison, G.W. Stuart, T.M. Wilkie, G. Norstedt, and R.D. Palmiter. 1984. Regulation, linkage, and sequence of mouse metallothionein I and II genes. *Mol. Cell. Biol.* **4**: 1221.

Seguin, C., B.K. Felber, A.D. Carter, and D.H. Hamer. 1984. Competition for cellular factors that activate metallothionein gene transcription. *Nature* **312**: 781.

Stuart, G.W., P.F. Searle, H.Y. Chen, R.L. Brinster, and R.D. Palmiter. 1984. A 12 base pair DNA motif that is repeated several times in metallothionein gene promoters confers metal regulation to a heterologous gene. *Proc. Natl. Acad. Sci.* **81**: 7318.

Yagle, M.K. and R.D. Palmiter. 1985. Coordinate regulation of metallothionein I and II genes by heavy metals and glucocorticoids. *Mol. Cell. Biol.* **5**: 291.

Regulation of Human β-Interferon Gene Expression by an Inducible Enhancer Element

S. Goodbourn, K. Zinn, and T. Maniatis

Department of Biochemistry and Molecular Biology, Harvard University
Cambridge, Massachusetts 02138

β-Interferon (β-IFN) is produced by fibroblasts in response to viral infection (for review, see Stewart 1979). In vitro, expression of β-IFN can be induced in fibroblast cell lines by both virus and double-stranded RNA. The induction is due, at least in part, to an increase in the number of RNA polymerase molecules on the gene. A number of laboratories have established systems in which cloned IFN genes are appropriately regulated after introduction into heterologous cells (for references, see Zinn et al. 1983). We have analyzed the *cis*-acting DNA sequence requirements for induction of the β-IFN gene by investigating the expression of in vitro–manipulated recombinants in mouse C127 fibroblasts. These recombinants are maintained on the extrachromosomal bovine papilloma virus (BPV) vector pBPV-BV1.

We have shown previously that the 5′ boundary of sequences required for the induction of expression of the human β-IFN gene by poly(I)-poly(C) lies between 77 bp and 73 bp upstream from the transcription initiation site (referred to hereafter as −77 and −73, respectively). In addition, an internal deletion of nucleotides −19 to +277 has no effect on the induction of β-IFN mRNA (Zinn et al. 1983). Thus, it seemed that sequences involved in β-IFN gene regulation could be localized to between −77 and −19. These results are consistent with those of Ohno and Taniguchi (1983), who find that a β-IFN fragment from −284 to +20 can confer inducibility by virus on the herpes simplex virus thymidine kinase (*tk*) structural gene.

To further delineate the sequences required for induction, we constructed a hybrid promoter in which β-IFN nucleotides from upstream of the TATA box (specifically −77 to −37) were fused to the

TATA box, mRNA cap site, and 5' untranslated region of the *tk* gene. This hybrid promoter was in turn fused to the coding and 3' noncoding sequences of the β-IFN gene. After treatment of transformed cell lines carrying this gene with poly(I)-poly(C), a very large increase in the levels of human β-IFN mRNA levels was observed. This induction is specific to the β-IFN −77/−37 region since no significant induction was observed when the β-IFN coding region and 3' noncoding region were fused to promoters that lacked these sequences. We have termed the β-IFN regulatory sequences that reside between −77 and −37 the *i*nterferon gene *r*egulatory *e*lement (IRE).

The IRE can act on the β-IFN gene and TATA box in an orientation-independent manner. In addition, we have also found that the IRE can confer inducibility when fused to a complete *tk* promoter in an orientation-independent manner. In this case transcription is initiated from the normal *tk* startpoint. We next determined whether the IRE retained its ability to confer inducibility on a linked promoter when the distance between these elements was increased. When the IRE was positioned 800 bp 5' to the *tk* promoter, a small, but reproducible, level of induction could still be observed. This demonstrates that the IRE has at least some of the properties of an inducible enhancer element. Unfortunately, the levels of mRNA from the *tk* promoter in these experiments are rather high before induction, and we reasoned that the induction ratio might be limited by saturating the maximal transcriptional capacity of this promoter. To further investigate the enhancer properties of the IRE, we therefore looked for a promoter with a lower basal activity. Since the IRE could function when fused in either orientation to a β-IFN 5'-deletion mutant that contains only the TATA box from the promoter region, we analyzed the activity of the IRE at a number of locations relative to the β-IFN TATA box (the −40 5'-deletion mutant of β-IFN). However, inducibility was only observed when the IRE was actually juxtaposed to the TATA box. These results are reminiscent of the properties of the SV40 enhancer, which is unable to enhance transcription from a truncated promoter unless it is adjacent (see Treisman and Maniatis 1985 for discussion). One explanation of this result could be that the −40 β-IFN truncation lacks upstream promoter sequences that are necessary to permit induction. To address this possibility, we repeated these experiments using a β-IFN truncation with 73 bp of 5'-flanking sequence. This mutant is only inducible ~5-fold by poly(I)-poly(C) and also has considerably lower basal activity than the highly inducible −77 deletion gene. When the IRE was placed up to 900 bp upstream of the −73 gene or 360 bp downstream of this gene, the levels of β-IFN mRNA could be induced at least 20-fold. Thus the IRE has all the properties of an inducible enhancer.

Only 14 bp of the IRE (−77 to −64) is required to cause an increase in the inducibility of the −73 deletion when these sequences are positioned 100 bp upstream. Because the 5′-deletion series we constructed does not have any mutants between −77 and −73, the 5′ end of this sequence could be even further 3′ to −77. A computer search for homology showed that the 13-bp sequence between −76 and −64 is related to a number of other sequences within the β-IFN gene. There are five of these sequence elements lying between position −108 and the mRNA cap site (Fig. 1). We believe that these sequence elements have functional importance for several reasons. First, as discussed above, a single copy of a 14-bp sequence, which completely contains the 13-bp element, can effectively "rescue" the low inducibility of a −73 β-IFN 5′-deletion mutant; the −73 breakpoint would inactivate one of these elements. Second, 5′-deletion mutants that include additional copies of this element are increasingly active in C127 cells. Third, differences in the 5′ boundary requirement are seen for different cell lines, and these correlate rather precisely with the junctions of the 13-bp elements. Whereas the 5′ boundary for efficient induction in C127 cells lies at the 5′ end of the third most cap-distal element, Fujita et al. (1985) find that in L929 cells sequences 3′ to −107 are essential for efficient induction; this boundary is at the 5′ end of the fifth most cap-distal element (Fig. 1). We also note that a −77 β-IFN 5′-deletion mutant is not inducible in HeLa cells. It is possible that the concentrations of a putative *trans*-acting factor are lower in L929 cells than in C127 cells, and therefore the additional copies of the 13-bp element present in the −107 mutant as compared with the −77 deletion would be important in maximizing the efficiency of promoter usage in different cells. Alternatively, the elements between −77 and −107 may interact with qualitatively different factors. However, the fact that these sequences are homologous suggests that such putative factors are likely to be similar.

It is also interesting to note that there are two sequences related to the 13-bp element 3′ to the gene as well as one weak homology within the gene itself. We have observed that a decrease in the induction ratio occurs if the body of the β-IFN gene used in these experiments is replaced by that of the human growth-hormone gene. This result suggests that the β-IFN transcription unit and the 3′-flanking sequences contain sequences that do not confer inducibility by themselves but augment the level of induction that can be

HUMAN β-IFN

Figure 1 Regulatory elements upstream from the human β-IFN gene. Sequences implicated in the induction of β-IFN mRNA synthesis of virus or poly(I)-poly(C) are shown. Numbers above the sequence indicate the number of base pairs upstream from the mRNA cap site, which is indicated by the arrow. The IRE located between −37 and −77 is shown, and the reiterated 13-bp regulatory sequences are shaded gray. These sequences are numbered I to V in a 5′ to 3′ direction.

attained from the IRE; we speculate that the 13-bp elements in the coding and 3'-flanking regions may be such sequences.

The presence of multiple reiterated elements within promoter and regulatory elements has been noted in other systems. In addition, many enhancer elements contain such sequence arrangements. It is possible that the presence of linked reiterated elements is important to facilitate binding of *trans*-acting factors by allowing cooperative binding. In this respect at least, enhancers may be similar to upstream promoter elements and regulatory sequences. The IRE is clearly related to both enhancers and regulatory promoter sequences. It is possible that the differences between such classes of element may be merely quantitative.

REFERENCES

Fujita, T., S. Ohno, H. Yasumitsu, and T. Taniguchi. 1985. Delimitation and properties of DNA sequences required for the regulated expression of human interferon-β gene. *Cell* **41:** 489.

Ohno, S. and T. Taniguchi. 1983. The 5' flanking sequence of human interferon-β1 gene is responsible for viral induction of transcription. *Nucleic Acids Res.* **11:** 5403.

Stewart II, W.E., ed. 1979. *The interferon system*. Springer-Verlag, New York.

Treisman, R. and T. Maniatis. 1985. Simian virus 40 enhancer increases the number of RNA polymerase II molecules on linked DNA. *Nature* **315:** 72.

Zinn, K., D. DiMaio, and T. Maniatis. 1983. Identification of two distinct regulatory regions adjacent to the human β-interferon gene in mouse cells. *Cell* **34:** 865.

Transcriptional Regulation of a Human Histone H4 Gene during the Cell Cycle

N. Heintz, G.C. Bleecker, O. Capasso, and S.M. Hanly

Laboratory of Biochemistry and Molecular Biology
The Rockefeller University, New York, New York 10021

To divide, the cell must replicate and distribute its DNA into two daughter cells (the chromosome cycle) and it must double its initial mass, including all of its structural and functional components (the growth cycle). Consideration of this problem has been facilitated by the concept of a cell cycle, which can be divided into four phases: G_1, S (DNA synthesis), G_2, and M (mitosis). Quiescent cells are sometimes considered to be a metabolically distinct state termed G_0. Although little is known about the control of entry into the cell cycle, it is quite clear that once the cell becomes committed to divide, it must then carry out a defined program that involves the production of a wide variety of molecules utilized chiefly during specific phases of the cell cycle. In many cases, the temporally regulated synthesis of these products reflects similar temporal expression of one, or several, specific genes. Although transformed cells are known to lack the normal mechanisms of growth control, they must still execute this program correctly to divide. Thus, we are interested in defining those mechanisms regulating the activity of specific genes during progression though the cell cycle and in discovering the processes that couple their expression to specific biochemical events.

Histone Gene Expression

The original observation that nuclear DNA synthesis and histone protein synthesis occur at significant rates only during the S phase of cultured HeLa cells (Robbins and Borun 1967) has resulted in a large number of studies concerning the coupling of these events (see Maxson et al. 1983). It has recently become evident that an accumulation of histone mRNA during S phase is the result of an approximately fivefold increase in both the rate of synthesis and the stability of histone mRNA during DNA synthesis (Heintz et al.

1983; Sittman et al. 1983). Furthermore, it was shown that histone gene expression can be modulated several times during a single cell cycle by repeated treatment with DNA synthesis inhibitors (Heintz et al. 1983) and that histone gene transcription is increased within 10 min after release from a block in DNA synthesis (Graves and Marzluff 1984). These observations demonstrate the dynamic nature of the regulation of mammalian histone gene transcription and suggest that it may involve a *trans*-acting regulatory factor of transcription.

One approach we have taken toward understanding this process is to attempt to reproduce these events in vitro. Our initial study in this area (Heintz and Roeder 1984) described the preparation of nuclear extracts from synchronized HeLa cells that appear to mimic the in vivo regulation of a human histone H4 gene. Hence, extracts from the S-phase cells transcribed this H4 gene more efficiently than those from non-S-phase cells. In contrast, the adenovirus major late promoter was most active in the non-S-phase extracts. Competition experiments indicated that the limiting factor for H4 transcription in vitro was specifically sequestered by preincubation with the H4 template DNA. To assess whether these results do, in fact, accurately reflect the in vivo situation, we have begun to examine the nucleotide sequences required for regulation in vitro and in vivo.

RESULTS AND DISCUSSION
The structure of the human histone H4 promoter and a series of deletion mutants employed in these studies are shown in Figure 1. Analysis of the transcription efficiency of these various mutants in vitro (Hanly et al. 1985) provided the following results:

1. Maximally efficient transcription of this gene in vitro requires distal promoter sequences in addition to the TATA box and cap site.
2. These distal promoter elements are recognized preferentially in S-phase nuclear extracts.
3. It is possible to restore to non-S-phase extracts the ability to recognize these distal sequences using fractions from S-phase cells.

These results further support the idea that histone gene regulation during the cell cycle involves a histone-specific transcription factor and suggest that this factor may act through upstream sequences of the H4 promoter. A definitive statement of this idea requires, of course, the direct demonstration that these sequences mediate cell-cycle regulation in vivo.

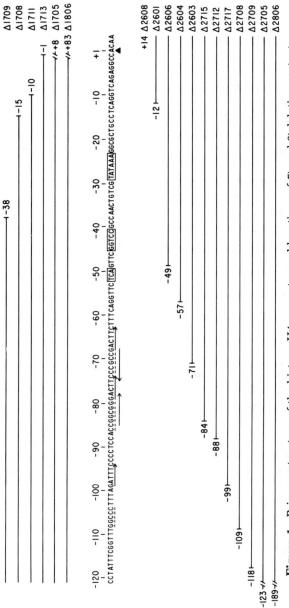

Figure 1 Primary structure of the histone H4 promoter and locations of 5'- and 3'-deletion mutants.

157

Preliminary experiments in which mutant H4 genes were introduced into mouse L cells and assayed during transient expression in vivo demonstrated that this human H4 gene is accurately transcribed in murine cells and that maximal expression in vivo is dependent on the H4 distal promoter elements. However, this type of transient assay cannot be employed to determine whether these sequences are sufficient for transcriptional regulation during S phase. Accordingly, we have constructed stable L-cell lines that carry the human H4 genes and are presently analyzing the expression of the introduced genes in synchronized cell populations. Although our analysis of these cell lines in not completed, our initial studies of several independently isolated cell lines containing the wild-type human H4 gene establish that the regulation of this gene in vivo requires an H4-specific *trans*-acting factor.

Seventeen mouse L-cell lines carrying the human H4 gene were prepared by cotransfection with the herpes virus *tk* gene and selected in HAT medium. Genomic Southern blot analysis of these cell lines showed that they contained between 1 and ~60 copies of the introduced H4 gene per haploid genome and that most of these copies were tandemly integrated. S1 nuclease analysis indicated that there was a general correlation between copy number and level of expression of the integrated human H4 genes. Furthermore, the steady-state levels of both the mouse and human H4 mRNA produced in several of these cell lines were cell-cycle regulated. In one interesting cell line containing 60 copies of the human H4 gene per haploid genome (Fig. 2), S1 analysis showed that only the human H4 mRNA was present at significant levels in the steady-state population. To determine whether integration of the human H4 genes had somehow rendered the mouse genes incapable of being expressed, we selected tk⁻ revertant cell lines in the presence of bromodeoxyuridine. Southern blot analysis demonstrated that the revertant cell lines had lost most or all of the human H4 genes, and S1 analysis indicated that the mouse genes are again expressed at normal levels. The resident mouse genes in clone 6-8, therefore, are fully functional but are not expressed, presumably due to competition with the introduced human H4 genes. Finally, S1 and Northern blot analysis demonstrated that the expression of both the mouse H3 and H2A genes remain unperturbed in this cell line. We conclude that mammalian histone H4 gene expression involves an H4-specific *trans*-acting factor.

To determine whether this factor could be responsible for the transcriptional regulation of the H4 genes, we have studied the rate of synthesis of H4 mRNA in this cell line. Nuclear run-on transcrip-

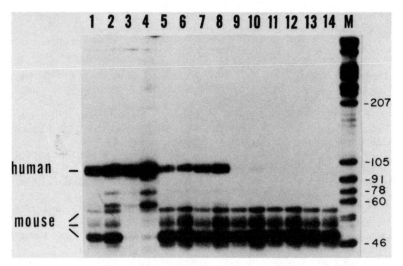

Figure 2 S1 nuclease mapping of total RNA from several transfected cell lines: (lanes *1* and *2*) clone 6-3; (lanes *3* and *4*) clone 6-8; (lanes *5* and *6*) clone 6-9; (lanes *7* and *8*) clone 8-2; (lanes *9* and *10*) clone 8-3; (lanes *11* and *12*) clone 8-4; (lanes *13* and *14*) Ltk⁻ parental cell line.

tion assays indicate that the rate of H4 mRNA synthesis in clone 6-8 is elevated at least 10-fold and perhaps as much as 50-fold over the parental Ltk⁻ cells. As expected, the rates of synthesis of the mouse H3 and H2A genes are unchanged. Finally, similar transcription studies showed that H4 transcription in clone 6-8 increases approximately 5-fold upon entry into S phase. These results very strongly suggest that the competition we observe in clone 6-8 is due to the presence of an H4-specific, *trans*-acting transcriptional regulatory protein. However, the extensive homology between the mouse and human H4 genes has so far precluded a direct demonstration that the mouse H4 genes in clone 6-8 are transcriptionally silent.

CONCLUSIONS

The in vivo results we have outlined above strongly support the results we have obtained concerning the regulation of histone gene transcription in vivo. It seems possible that the factor that interacts with the H4 distal transcription elements in vitro is identical with the factor that is responsible for the in vivo competition observed in clone 6-8. Demonstration of this identity will require precise localization of the sequences responsible for histone gene regulation in vivo and the isolation and characterization of the H4-specific transcription factor.

REFERENCES

Graves, R.A. and W.F. Marzluff. 1984. Rapid reversible changes in the rate of histone gene transcription and histone mRNA levels in mouse myeloma cells. *Mol. Cell. Biol.* **4:** 351.

Hanly, S.M., G.C. Bleecker, and N. Heintz. 1985. Identification of promoter elements necessary for transcriptional regulation of a human histone H4 gene in vitro. *Mol. Cell. Biol.* **5:** 380.

Heintz, N. and R.G. Roeder. 1984. Transcription of human histone genes in extracts from synchronized HeLa cells. *Proc. Natl. Acad. Sci.* **81:** 1713.

Heintz, N., H.L. Sive, and R.G. Roeder. 1983. Regulation of human histone gene expression: Kinetics of accumulation and changes in the rate of synthesis and in the half lives of individual histone mRNAs during the HeLa cell cycle. *Mol. Cell. Biol.* **3:** 539.

Maxson, R., R. Cohn, L. Kedes, and T. Mohun. 1983. Expression and organization of histone genes. *Annu. Rev. Biochem.* **17:** 239.

Robbins, E., and T.W. Borun. 1967. The cytoplasmic synthesis of histones in HeLa cells and its temporal relationship to DNA replication. *Proc. Natl. Acad. Sci.* **58:** 1977.

Sittman, D.B., R.A. Graves, and W.F. Marzluff. 1983. Histone mRNA concentrations are regulated at the level of transcription and mRNA degradation. *Proc. Natl. Acad. Sci.* **80:** 1849.

Evidence for a Common Program of Proto-oncogene Expression Induced by Diverse Growth Factors

M.E. Greenberg,* L.A. Greene †, and E.B. Ziff*

*Department of Biochemistry and Kaplan Cancer Center
and †Department of Pharmacology
New York University Medical Center, New York, New York 10016

In higher eukaryotes, polypeptide growth factors are important regulators of cellular proliferation and differentiation (Carpenter and Cohen 1979; Greene and Shooter 1980; Stiles 1983). Growth factors bind to specific membrane-associated receptors that, in turn, transduce signals for proliferation and differentiation to the cell nucleus. In this paper we summarize evidence from our laboratories that a variety of different growth factors, each of which interacts with a particular membrane-associated receptor present in distinct cell types, induce very similar changes in the transcription of specific cellular genes. Because these changes in gene expression are rapid, occurring within minutes of growth-factor binding, we propose that they are primary features of the mechanism of growth-factor action. Significantly, certain of the genes induced by growth factors are proto-oncogenes (for review, see Bishop 1983). Thus, the transcriptional program induced in diverse cell types by growth factors appears to regulate genes that are basic to the control of cell proliferation and differentiation. Together, these results provide evidence for a common program of gene control that serves a variety of cell types during proliferation and differentiation and is stimulated by a series of otherwise distinct growth factor–receptor interactions.

Assay for Gene Regulation by Growth Factors

Our experiments started with the assumption that growth factors regulate the transcription of specific cellular genes. We utilized a nuclear run-off transcription assay to identify the proposed growth-factor-regulated genes (Greenberg and Ziff 1984a). Initially, this assay was applied to changes in gene transcription that occur when quiescent BALB/c-3T3 cells are stimulated with serum to progress from the G_0 resting state into the G_1 growth state. Serum or purified growth factors were added to monolayer cultures of resting cells, the cells were harvested at various times poststimulation, and nuclei were isolated and incubated with ^{32}P-labeled nucleotide triphosphates to yield highly radioactive "run-off" RNA products. This run-off RNA is representative of the nuclear transcripts that are in the process of being synthesized at the time of cell harvest. Run-off RNA was hybridized to a series of cDNA clones immobilized on nitrocellulose filters. The level of hybridization is visualized by autoradiography and provides a measure of the transcriptional activity of individual genes. In this manner, the level of transcription of large numbers (>30) of genes is simultaneously analyzed as a function of time after growth-factor stimulation.

Growth-factor Stimulation of Proto-oncogene Transcription
In our initial studies we monitored the transcriptional activity of a panel of proto-oncogenes (Greenberg and Ziff 1984a). Abnormal expression of the proto-oncogenes can cause unregulated cellular proliferation (for review, see Bishop 1983), raising the possibility that some members of the proto-oncogene family might participate directly in the control of cell growth and that their expression might be regulated by growth factors.

When quiescent 3T3 cells were stimulated by fresh medium containing 15% serum, the level of transcription of the c-*fos* proto-oncogene rapidly increased greater than 15-fold as shown in Figure 1. c-*fos* encodes a nuclear protein of unknown function whose elevated expression in mutant form can transform cells (for review, see Verma 1984). Transcription of actin sequences was also rapidly stimulated. The activation of c-*fos* expression was detected within 5 min and reached a maximum at approximately 10 min poststimulation. This is the most rapid transcriptional response occurring in animal cells stimulated by peptide growth hormones yet reported. The rapid kinetics of the activation suggest that the transcriptional increase is a direct effect of growth-factor action.

At approximately 30 min poststimulation, the level of c-*myc* transcription is also increased. c-*myc* encodes a nuclear protein whose abnormal expression is implicated in a wide range of tumors (for review, see Leder et al. 1983). The delay in the increase in c-*myc* transcription rate relative to c-*fos* and actin is a consistent feature of growth-factor activation (see also below), and this fact must be accounted for by any model for the activation mechanism.

In these experiments, cytoplasmic c-*fos* and c-*myc* mRNAs, analyzed by Northern blotting, showed a dramatic increase directly

162

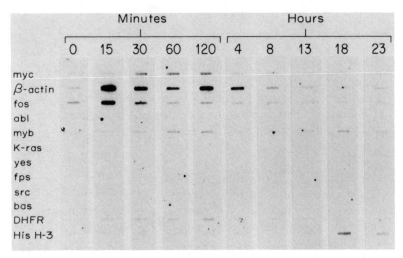

Figure 1 Kinetics of the transcription of a series of cellular genes when quiescent 3T3 cells are stimulated by serum to enter the G_1 growth state and proceed through the cell cycle. Reprinted, with permission, from Greenberg and Ziff (1984b).

after the elevation of transcription levels in the nucleus (Greenberg and Ziff 1984a). This increase was transient. In the case of c-*fos*, the mRNA level declined to baseline by 30 min poststimulation.

We have also shown that purified growth factors can activate the transcription of c-*fos*, c-*myc*, and actin in quiescent 3T3 cells. Both platelet-derived growth factor (PDGF) and fibroblast growth factor (FGF) induced the transcription of these genes. Treatment with the phorbol ester TPA also generates the same pattern of transcriptional activation (Greenberg and Ziff 1984a). When the kinetics of activation of c-*fos* transcription by PDGF and TPA were compared, they were found to be virtually identical.

Proto-oncogene Regulation in PC12 Cells by Nerve Growth Factor and Epidermal Growth Factor

We next asked whether the regulation of c-*fos*, c-*myc*, and actin transcription was a special feature of the effects of growth factors upon quiescent 3T3 cells, or whether other growth factors acting on quite different cell types could also activate the expression of these genes. For this purpose we have analyzed nerve growth factor (NGF)–responsive transcription in a transformed chromaffin cell line, PC12, derived from a rat adrenal medullary pheochromocytoma (Greene and Tischler 1976). In tissue-culture medium supple-

163

mented only by serum, PC12 cells grow slowly and possess a round morphology typical of chromaffin cells. The addition of the hormone NGF causes the cessation of PC12 cell division and is accompanied by neuronal differentiation, including outgrowth of neuronal processes (for review, see Greene and Tischler 1982). The full effects of NGF require a period of up to 14 days to take place. The NGF-induced changes recapitulate many of the features of in vivo neuronal differentiation. PC12 cells provide an excellent in vitro model for studying the mechanism of action of NGF during neuronal differentiation.

We have analyzed the effects of NGF action on the transcription of c-*fos*, c-*myc*, and actin sequences in PC12 cells (M. Greenberg et al., in prep.). We find that NGF stimulates the transcription of these genes with kinetics that directly parallel the changes first observed in fibroblasts when induced by PDGF. Epidermal growth factor (EGF) is mitogenic for PC12 cells but does not stimulate neuronal differentiation (for review, see Greene and Tischler 1982). We have found that EGF also stimulates in PC12 cells the rapid activation of c-*fos*, actin, and c-*myc* transcription (M. Greenberg et al., in prep.). Thus, PDGF action on fibroblasts as well as NGF and EGF action on PC12 cells stimulate similar expression of two proto-oncogenes and actin. The activation of early-response genes by these and other factors is summarized in Table 1.

These results suggest that the induction of c-*fos*, c-*myc*, and actin may be a general nuclear response to growth or differentiation factors acting through different membrane-associated receptors in distinct cell types and that the proteins encoded by these genes may mediate the subsequent cellular responses to growth factors.

Table 1
Activation of Early-response Genes

	BALB/c-3T3	PC12
PDGF	+ + +	−
NGF (50 ng/ml)	−	+ + +
EGF (3 ng/ml)	+	+ + +
FGF	+ + +	+
TPA (50 nM)	+ + +	+ +
insulin (5 μg/ml)	−	+
db cAMP (1 mM)	?	+
KCl (46 mM)	?	+

ACKNOWLEDGMENTS
This work was supported by American Cancer Society grant MV-75 and National Institutes of Health grant GM-30760 to E.B.Z. and NS-16036 to L.A.G. E.B.Z. is the recipient of ACS Faculty Research Award FRA 207. M.E.G. was the recipient of a postdoctoral fellowship from the Damon Runyon–Walter Winchell Cancer Fund and a postdoctoral fellowship from the U.S. Public Health Service.

REFERENCES

Bishop, J.M. 1983. Cellular oncogenes and retroviruses. *Annu. Rev. Biochem.* **52:** 301.

Carpenter, G. and S. Cohen. 1979. Epidermal growth factor. *Annu. Rev. Biochem.* **48:** 193.

Greenberg, M. and E. Ziff. 1984a. The stimulation of 3T3 cells induces transcription of the c-*fos* proto-oncogene. *Nature* **311:** 433.

———. 1984b. c-*fos* transcription is activated as an early response to 3T3 cell stimulation. *Cancer Cells* **3:** 307.

Greene, L.A. and E.M. Shooter. 1980. The nerve growth factor: Biochemistry, synthesis, and mechanism of action. *Annu. Rev. Neurosci.* **3:** 353.

Greene, L.A. and A.S. Tischler. 1976. Establishment of a noradrenergic clonal line of rat adrenal pheochromocytoma cells which respond to nerve growth factor. *Proc. Natl. Acad. Sci.* **73:** 2424.

———. 1982. PC12 pheochromocytoma cultures in neurobiological research. *Adv. Cell. Neurobiol.* **3:** 31373.

Leder, P., J. Battey, G. Lenoir, C. Moulding, W. Murphy, H. Potter, T. Stewart, and R. Taub. 1983. Translocations among antibody genes in human cancer. *Science* **22:** 771.

Stiles, C.D. 1983. The molecular biology of platelet-derived growth factor. *Cell* **33:** 653.

Verma, I.M. 1984. From c-*fos* to v-*fos*. *Nature* **308:** 317.

Regulated Transcription of a Cloned c-*fos* Gene in Response to Serum Factors

R. Treisman

MRC Laboratory for Molecular Biology
Cambridge CB2 2QH, England

The c-*fos* gene, cellular homolog of the oncogene carried by the FBJ murine osteosarcoma virus (Finkel et al. 1966; Curran et al. 1983), provides an interesting example of regulated gene expression. The gene is expressed as a 2.2-kb RNA encoding a 55-kD protein, p55[c-*fos*], and is expressed at a high level in extraembryonal tissues (Müller et al. 1983). The function of the gene is unknown, but it is thought to be involved in pathways of cellular differentiation (Müller and Wagner 1984; Mitchell et al. 1985). Recently, it was reported that the c-*fos* gene exhibits a novel pattern of expression when quiescent fibroblasts are stimulated by whole serum or purified growth factors such as platelet-derived growth factor (PDGF) that can induce competence for DNA replication (Greenberg and Ziff 1984; Kruijer et al. 1984; Müller et al. 1984). High rates of transcription are detected within 5 min of stimulation (Greenberg and Ziff 1984) and are accompanied by the accumulation of c-*fos* mRNA (Greenberg and Ziff 1984; Kruijer et al. 1984; Müller et al. 1984). However, within 30 min transcription rates return to prestimulation levels (Greenberg and Ziff 1984); c-*fos* mRNA is rapidly degraded and is essentially undetectable 2 hr after stimulation (Greenberg and Ziff 1984; Kruijer et al. 1984; Müller et al. 1984). When quiescent cells are treated with protein synthesis inhibitors such as cycloheximide, c-*fos* RNA levels are increased slightly. However, stimulation with both serum and inhibitor results in both greater c-*fos* mRNA accumulation and its long-term persistence (Müller et al. 1984), implicating short-lived proteins in transcription and/or mRNA turnover. Similar transcriptional changes are observed with BALB/c-3T3 (Greenberg and Ziff 1984), NIH-3T3 (Kruijer et al. 1984; Müller et al. 1984), and HeLa cells.

To study the molecular basis of c-*fos* gene regulation, a transient DNA transfection assay was used. Plasmids carrying the human c-*fos* gene were introduced into NIH-3T3 cells by the calcium phosphate coprecipitation technique. After 12–15 hr exposure to the precipitate, the serum-containing medium was replaced with medium containing 0.5% fetal calf serum, and the cells were left an additional 30–36 hr before stimulation with medium containing 15% fetal calf serum. A plasmid carrying the human α_1-globin gene was cotransfected to provide an internal control for transfection efficiency and RNA integrity. For RNA analysis, protection of a complementary strand RNA probe from nuclease digestion was measured. For preliminary experiments, the 5.4-kb BamHI fragment of human DNA containing the c-fos gene together with 0.75 kb of 5'-flanking sequences (Curran et al. 1983; a gift from I.M. Verma) was subcloned into pUC12. Analysis of the RNA produced by this plasmid over a 4-hr time course showed that a 20-fold to 50-fold increase in c-fos mRNA occurred within 30 min of serum stimulation. Unspliced transcripts were detectable no later than 30 min after stimulation, suggesting that transcription shut-off of the transfected gene occurred. Turnover of the human c-fos mRNA was detected in transfected 3T3 cells but appeared to be slower than that of the endogenous mouse c-fos mRNA. When transfected cells were treated with 10 μg/ml cycloheximide in addition to serum, both increased induction and continued persistence of c-fos mRNA were observed. These results suggest that the transient transfection assay provides a suitable system with which to study c-fos transcription regulation in response to serum factors.

To localize sequences essential for induction of c-fos transcription in response to serum factors, deletion derivatives of the gene were constructed and analyzed. The first such series comprised truncations of the gene possessing different amounts of 5'-flanking sequence. This analysis showed that sequences between −306 and −332 bp 5' to the mRNA cap site are essential for high-level induction (Table 1). To localize the 3' boundary of this putative 5' element, two series of internal deletions were constructed. The first set lacks sequences between a 5' endpoint and a 3' endpoint at −222 (5' deletion to −222 destroys high-level inducibility of transcription). Analysis of these deletions showed that sequences between −276 and −327 are required for efficient induction of transcription (Table 1). This result also suggests that no sequences 5' to −332 are involved in the regulation and that no sequences between −276 and −222 are necessary for inducibility. A second series of internal deletions combined the same 5' endpoints with a common 3' endpoint at −123. Removal of sequences between −222 and −123 did not significantly lower inducibility, suggesting that no essential sequence elements are located between these positions. Inspection of the

Table 1

Properties of Deletion Mutants of the Human
c-*fos* Gene in the Transient Serum Stimulation
Assay

Deletion	Efficient induction
5′ Deletion endpoint	
−711	+
−554	+
−484	+
−363	+
−332	+
−306	−
−290	−
−261	−
−222	−
Internal deletion	
−223/−262	+
−223/−276	+
−223/−327	−
−223/−349	−
−124/−262	+
−124/−276	+
−124/−327	−
−124/−349	−

Deletions are given relative to the mRNA cap site.
RNA levels were measured at 30 min after addition of
serum-containing medium. Efficient induction represents a 20-fold to 50-fold increase in c-*fos* mRNA level
during this period.

DNA sequence of the c-*fos* gene between −276 and −327 reveals several patches of sequence homology to the enhancers from polyoma (cf. c-*fos* nucleotides −287 to −327 with polyoma nucleotides 5117 to 5154) and SV40 (cf. c-*fos* nucleotides −271 to −304 with SV40 nucleotides 174 to 141). The significance of these homologies is under investigation.

The properties of the c-*fos* 5′-flanking sequences were further examined. For these experiments a segment of c-*fos* DNA including nucleotides −711 to −120, attached to synthetic *Eco*RI linkers, was used. Inversion of this segment has little effect, if any, on inducibility. The segment was inserted into the −222 5′-deletion mutant in either orientation at position −222, 100 bp 5′ to its normal position. Normal induction of transcription was observed in both cases. When the upstream segment was inserted at the 3′ side of the gene, more than 2500 bp distant, inducibility was also restored; however, no more than 5-fold induction was observed in this case. These results suggest that although the c-*fos* upstream element can function in an orientation-independent manner, its function is strongly distance-dependent. When the SV40 enhancer element was inserted at these positions, constitutive synthesis of c-*fos* RNA was observed, regardless of the 5' or 3' position of the SV40 sequences.

To investigate the functional role of c-*fos* 5'-flanking sequences, fusion genes were constructed, comprising c-*fos* nucleotides −711 to +42 joined to heterologous transcription units. One such fusion gene, c-*fos*/*cat*, contains the c-*fos* 5'-flanking sequences joined to the chloramphenicol acetyltransferase gene from pSV2CAT. In the transient transfection assay, serum-dependent induction of c-*fos*/*cat* RNA synthesis was observed, indicating that the c-*fos* 5'-flanking sequences can confer regulation on a heterologous transcription unit. The kinetics of induction in this case, however, appeared different from the intact c-*fos* gene, with RNA accumulating up to 4 hr poststimulation. The reasons for this observation are unclear; possibly further regulatory elements within the c-*fos* gene itself are disrupted in the c-*fos*/*cat* fusion gene. Experiments with a fusion gene in which the c-*fos* 5'-flanking sequences are joined to the human β-globin transcription unit are consistent with this hypothesis. To test whether elements close to the c-*fos* mRNA cap site are also necessary for inducibility of c-*fos* transcription, sequences from −119 to +49 were excised from the c-*fos* gene and replaced with β-globin sequences from −128 to +20. The results indicate that this mosaic gene remains inducible in the transient transfection assay. A β/c-*fos* fusion gene lacking the c-*fos* upstream sequences is no longer inducible. These results suggest that no sequences within the immediate 5'-flanking region of c-*fos* are necessary for serum inducibility and that at least part of the serum response is mediated by the c-*fos* upstream sequences.

SUMMARY

Sequences in the 5'-flanking region of the human c-*fos* gene are required for the correctly regulated response of the gene to serum factors. The 5' border of these sequences is located between nucleotides −332 and −306 and the 3' border between nucleotides −276 and −327 relative to the mRNA cap site. There appears to be no essential element for serum induction between nucleotides −119 and +49, since these nucleotides can be replaced with analogous sequences from the human β-globin gene. Work is in progress to characterize further the upstream sequence element and to determine whether other sequences are involved in regulated c-*fos* gene expression.

REFERENCES

Curran, T., W.P. MacConnell, F. van Straaten, and I. Verma. 1983. Structure of the FBJ murine osteosarcoma virus genome: Molecular cloning of its associated helper virus and the cellular homolog of the v-*fos* gene from mouse and human cells. *Mol. Cell. Biol.* **3:** 914.

Finkel, M.P., B.O. Biskis, and P.B. Jinkins. 1966. Virus induction of osteosarcomas in mice. *Science* **151:** 698.

Greenberg, M.E. and E.B. Ziff. 1984. Stimulation of 3T3 cells induces transcription of the c-*fos* proto-oncogene. *Nature* **311:** 433.

Kruijer, W., J.A. Cooper, T. Hunter, and I.M. Verma. 1984. PDGF induces rapid but transient expression of the c-*fos* gene. *Nature* **312:** 711.

Mitchell, R.L., L. Zokas, R.D. Schreiber, and I.M. Verma. 1985. Rapid induction of the proto-oncogene *fos* during human monocytic differentiation. *Cell* **40:** 209.

Müller, R. and E.F. Wagner. 1984. Differentiation of F9 teratocarcinoma stem cells after transfer of c-*fos* proto-oncogenes. *Nature* **311:** 438.

Müller, R., I. Verma, and E.D. Adamson. 1983. Expression of c-*onc* genes: c-*fos* transcripts accumulate to high levels during development of mouse placenta, yolk sac and amnion. *EMBO J.* **2:** 679.

Müller, R., R. Bravo, J. Burckhardt, and T. Curran. 1984. Induction of c-*fos* gene and protein by growth factors precedes activation of c-*myc*. *Nature* **312:** 716.

Yeast UASs as Coordinators of Gene Expression

L. Guarente

Department of Biology, Massachusetts Institute of Technology
Cambridge, Massachusetts 02139

Sites ostensibly similar to mammalian enhancers have been found in the yeast *Saccharomyces cerevisiae*; these sites, which are termed upstream activation sites (UASs), mediate regulation of transcription in accordance with growth conditions or physiology (Guarente 1984). The presence in *cis* of UASs is an absolute requirement for in vivo transcription of genes by RNA polymerase II under inducing conditions. Well-characterized UASs include those of the *GAL1–10* gene cluster (Guarente et al. 1982; West et al. 1984), the *HIS3* and *HIS4* loci (Struhl 1982; Hinnebusch et al. 1985), and the subject of this report, the *CYC1* gene, encoding the iso-1-cytochrome *c*. The iso-1 form makes up about 90% of the cytochrome *c* in yeast. The minor form, iso-2, is encoded by the *CYC7* locus.

CYC7

Previous experiments employing *CYC1-lacZ* fusions indicated that *CYC1* transcription was induced 200-fold when a heme supplement was added to cells growing under heme-deficient conditions. A further 10-fold derepression could be achieved by shifting these cells from glucose-containing media to media containing a nonfermentable carbon source (Guarente and Mason 1983). Regulation by both heme and carbon source was mediated by either of two UASs that lie at −275 (UAS1) and −225 (UAS2) in the *CYC1* upstream region (Guarente et al. 1984) (Fig. 1). When both sites are present, their activation of transcription in response to heme and carbon source is additive. UAS1 and UAS2 bear a core of homology with a match of 13 out of 16 bp.

TATA Box

The *CYC1* promoter contains a second important region bearing several potential TATA boxes and seven mRNA initiation sites spanning a 35-bp region. Deletion of the most upstream TATA box reduces levels of transcription about 5-fold, while deletion of the sec-

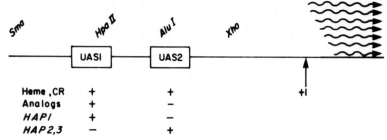

Figure 1 The *CYC1* promoter region. Wavy arrows represent *CYC1* transcripts. The upstream activation sites UAS1 and UAS2 are boxed. Plus indicates regulation, and minus indicates the lack of a regulatory response. CR stands for catabolite repression.

ondary TATA boxes or deletion of most of the initiation sites themselves exerts less than a 2-fold reduction in expression. Preliminary analysis of the 5′ ends of the mRNA in the deletion mutants suggests that although the TATA box determines levels of RNA, it does not fix the startpoint of transcription. It is believed that the region around the 5′ end itself encodes the mRNA start for *CYC1* and, possibly, in yeast in general.

UAS1 and UAS2

Employing a template with gaps at random locations across UAS1, misincorporation of α-S-dNTPs was carried out in vitro, and ten single-point mutations were introduced into UAS1. Nine of the ten mutations are defective in activation. Surprisingly, these nine changes do not lie in the core region of homology with UAS2 but in the region on either side of the core. Six mutations lie in a pentameric sequence ACCGA that lies about 15 bp upstream of the core. These mutations all exhibit a severely defective phenotype (10-fold down). Three other mutations are split between two other pentameric boxes similar to the ACCGA. One box, ACCGG, lies 5 bp upstream of the core, while the other, ACGGA, lies 5 bp downstream of the core. Mutations in the fourth homologous pentamer, ACCGA, lying 15 bp downstream of the core were not uncovered. Interestingly, all four pentameric boxes lie on one face of the helix and sandwich the core. The tenth mutation lies in the core and is silent phenotypically. The above mutations clearly indicate the importance of the pentamers in UAS1 activity but leave open the possibility that the core is not functional. Two experiments show that, in fact, the core is also a key determinant of UAS1 activity. First, a

12-bp deletion within the core reduces the activity of UAS1 5-fold to 10-fold. Second, a synthetic site bearing the pentamers with the spacing relationships as in the wild type but with the core substituted with an altered sequence was extremely defective for activity (about 100-fold down from the wild-type UAS1).

Our current view of UAS1 is thus of a site with two important components, the pentamers and the core. Since the site bearing the 12-bp deletion of the core still responds (albeit at a reduced level) to heme levels and to the *trans*-acting activator described below, we conclude that the pentamers mediate activation. A protein binding to the core may be required to promote binding of the activators on either side, perhaps by cooperative interactions with the activators. This requirement may be reduced if the flanking repeats are brought sufficiently close so that cooperative interactions between bound activators are made possible. This model would explain the partial activity of the site bearing a 12-bp deletion of the core. One appealing aspect of the model is that a wider range of regulation may be possible if two distinct proteins must bind to activate the site.

Much less information is available regarding important sequences in UAS2. Inspection of the DNA sequence of that site shows that UAS2 may be organized in a manner much like UAS1, with the core flanked by single direct repeats of the pentameric sequence CAAGC. Deletion of one pentamer leads to partial inactivation of UAS2. Experiments aimed at obtaining point mutations to verify the role of the pentamers in UAS2 are in progress.

Genes That Regulate UAS1 and UAS2

The above picture of essential elements in UAS1 and UAS2 makes an important prediction. *trans*-acting regulatory proteins should fall into two classes: those affecting both UAS1 and UAS2, and those specific for one site or the other. In yeast, such factors are most easily identified by regulatory mutations that act in *trans* to affect gene expression. It has been possible to obtain regulatory mutations affecting *CYC1* that map in three loci, *HAP1*, *HAP2*, and *HAP3*. Mutations in *HAP1* abolish the activity of UAS1 but not UAS2, whereas mutations in either *HAP2* or *HAP3* have just the opposite effect. Our working hypothesis is that the product of *HAP1* binds to the UAS1 pentamers, while a complex between the *HAP2* and *HAP3* products binds to the UAS2 pentamers to activate transcription.

To begin exploring the mechanism of *HAP*-mediated activation of the UAS sites, we are cloning these loci. So far, work on *HAP2* indicates that the locus encodes a 1.3-kb transcript that itself is regulated by catabolite repression. Furthermore, the product of *HAP2* appears to be localized to the yeast nucleus, consistent with the hypothesis that it is a direct activator of transcription. This point was demonstrated by constructing a *HAP2-lacZ* fused gene that retained *both* activities in vivo. Fixed cells were then treated with antibody to β-galactosidase, followed by rhodamine-conjugated IgG to localize the hybrid protein.

Coordination of *CYC1* Transcription with That of Other Genes Involved in Electron Transport

A comparison of *CYC1* regulation to that of other regulated genes may provide insight into global control systems operating in a eukaryotic cell and shed light on molecular mechanisms by which transcription is controlled. Examination of the phenotype of strains bearing *HAP1*, *HAP2*, or *HAP3* mutations immediately suggests that the *HAP1* system is cytochrome *c*–specific, while the *HAP2–3* system is global. A reduction in levels of all cytochromes is observed in *hap2* or *hap3* mutant strains, and such strains do not grow on media containing nonfermentable carbon sources.

Studies of the regulation of *CYC7* and *HEM1* (encoding the first step in the biosynthesis of the cytochrome cofactor, heme) have been initiated. Transcription of *CYC7* differs from that of *CYC1* in two respects. First, levels of *CYC7* transcription are about 5–10% of those of *CYC1*, and second, *CYC7* transcription is not turned off in the absence of heme. By employing *CYC7-lacZ* fusions, we have concluded that *CYC7* has an essential TATA box and a UAS region containing two or three discrete sites (T. Prezant and L. Guarente, unpubl.). Surprisingly, *CYC7* transcription is reduced 3-fold to 5-fold in *hap1* (but not *hap2* or *hap3*) mutants. Thus the *HAP1* product whose activity at UAS1 requires heme appears to activate the heme-insensitive *CYC7* gene.

Since the *HEM1* product catalyzes the committed step in heme biosynthesis, it is reasonable to posit that this gene would be involved in any coordination in the regulation of cytochrome and heme synthesis. To test this possibility, we have cloned *HEM1*, constructed *lacZ* fusions, and studied their regulation. It is clear that *HEM1* is activated by the global *HAP2-HAP3* system but not by *HAP1* (T. Keng and L. Guarente, unpubl.). A dilemma similar to that described above in the case of *CYC7* is posed by the finding that although *HEM1* falls under the *HAP2–3* umbrella, its synthesis is not reduced under conditions of heme deficiency. In fact, at high intracellular levels of heme, *HEM1* expression is repressed. Current experiments are aimed at unraveling this paradox.

DISCUSSION

The studies of *CYC1* regulation have provided several insights into transcription initiation in yeast. First, regulated promoters are modular, containing one or more regulated UASs and a TATA box–mRNA initiation region. The former sites are activated by specific regulatory proteins in response to gene-specific physiological signals. The TATA box determines levels of transcription, but where transcription initiates is encoded by the initiation sites themselves. In the case of the *CYC1* site, UAS1, it appears that the minimal site required for activation consists of two components, neither of which has activity alone. One element consists of a core sequence also found in UAS2, while the second element is a series of four pentameric repeats that sandwich the core. The repeats themselves encode recognition for the *trans*-acting activator, while the core, by hypothesis, encodes recognition for a second factor whose binding aids binding of the activator.

trans-acting mutations define activators that discriminate between UAS1 and UAS2. The product of one of these genes, *HAP2*, was shown to be nuclear, which is consistent with the hypothesis that the HAP proteins are the direct activators of transcription. The *hap⁻* strains also allow *CYC1* regulation to be integrated into the whole of control of functions involved in respiration in yeast. Results so far indicate that the *HAP1* system, which activates UAS1, also activates the *CYC7* gene encoding the minor form of cytochrome *c* but does not regulate other genes. In contrast, the *HAP2–HAP3* system, which activates UAS2, is global and regulates many, if not all, cytochrome genes, as well as the *HEM1* gene, which encodes the first step in heme biosynthesis. Future experiments should sharpen the picture of how regulation of this wide set of genes occurs at the molecular level.

ACKNOWLEDGMENTS

Work performed in the author's laboratory was supported by grants to L.G. by the National Institutes of Health (grant 5-R01-GM30454) and the W.R. Grace Company (research grant administered by M.I.T.).

REFERENCES

Guarente, L. 1984. Yeast promoters: Positive and negative elements. *Cell* **36:** 799.

Guarente, L. and T. Mason. 1983. Heme regulates transcription of the *CYC1* gene of *S. cerevisiae* via an upstream activation site. *Cell* **32:** 1279.

Guarente, L., R. Yocum, and P. Gifford. 1982. A *GAL10-CYC1* hybrid promoter identifies the *GAL4* regulatory region as an upstream site. *Proc. Natl. Acad. Sci.* **79:** 7410.

Guarente, L., B. Lalonde, P. Gifford, and E. Alani. 1984. Distinctly regulated tandem upstream activation sites mediate catabolite repression of the *CYC1* gene of *S. cerevisiae*. *Cell* **36:** 503.

Hinnebusch, A., G. Lucchini, and G. Fink. 1985. A synthetic *HIS4* regulatory element confers general amino acid control on the cytochrome *c* gene (*CYC1*) of yeast. *Proc. Natl. Acad. Sci.* **82:** 498.

Struhl, K. 1982. Regulatory sites for *HIS3* gene expression in yeast. *Nature* **300:** 284.

West, R., R. Yocum, and M. Ptashne. 1984. *S. cerevisiae GAL1-GAL10* divergent promoter region: Location and function of the upstream activating sequence UAS_G. *Mol. Cell. Biol.* **4:** 2467.

GAL4 – A Positive Regulator of Gene Activity in Yeast

M. Ptashne, R. Brent, E. Giniger, L. Keegan, and S. Varnum

Department of Biochemistry and Molecular Biology, Harvard University
Cambridge, Massachusetts 01238

When cells of the baker's yeast, *Saccharomyces cerevisiae*, are grown on galactose, transcription of the genes involved in galactose metabolism is induced more than 5000-fold. For the *GAL1* and *GAL10* genes, this induction depends on the upstream activating sequence (UAS$_G$), which is located midway between these divergently transcribed genes, and on the positive regulatory protein GAL4. Transcription of the *GAL* genes is repressed when cells are grown on glucose and can also be repressed by the specific, negative regulatory protein GAL80, which inhibits the action of GAL4 protein when cells are grown in the absence of galactose. We wish to understand how GAL4 protein activates *GAL* transcription and also how this activation is regulated. To this end, we have been analyzing the DNA-binding properties of GAL4 protein in vivo and in vitro and attempting to alter the pattern of *GAL* regulation by replacing some of the elements of the *GAL* regulatory circuit with genetic regulatory proteins from *Escherichia coli*.

DNA-binding of GAL4 Protein In Vivo

We have shown that GAL4 protein binds in vivo to four sites in UAS$_G$ to activate transcription of the *GAL1* and *GAL10* genes (Giniger et al. 1985). GAL4 protein expressed in *E. coli* protects guanine residues in UAS$_G$ from methylation by dimethyl sulfate. The same set of protections is seen in vivo in yeast and depends on the *GAL4*[+] allele. This protection pattern is consistent with the idea that GAL4 protein binds to four related 17-bp sequences, each of which displays approximate twofold rotational symmetry. A single near-consensus 17-bp oligonucleotide, installed in front of the yeast *GAL1* or *CYC1* transcription units, confers a high level of galactose inducibility upon these genes. Further experiments suggest that one mechanism of glucose repression is inhibition of the binding of GAL4 protein to DNA, but that *GAL80*-mediated repression of the

GAL genes does not greatly reduce the binding of GAL4 protein to UAS$_G$.

DNA-binding of GAL4 Protein In Vitro
In vitro, GAL4 protein produced in *E. coli* binds specifically to UAS$_G$ or to the 17-bp synthetic recognition sequence described above, as assayed by retention of the DNA on nitrocellulose filters. A hybrid protein bearing the aminoterminal 147 amino acids of GAL4 protein fused to *E. coli* β-galactosidase also binds specifically in this assay.

Positive Control by a LexA/GAL4 Hybrid Protein
A hybrid protein bearing the aminoterminal domain (87 residues) of the *E. coli* LexA protein fused to amino acid residues 74–881 of GAL4 protein functions both in *E. coli* and in yeast. In *E. coli*, it renders the cell sensitive to ultraviolet light, indicating that it binds to *lex* operators but cannot be efficiently inactivated by the *E. coli* SOS system. In yeast, the hybrid protein stimulates transcription of *GAL1* and *CYC1* if those genes bear a single *lex* operator 250–850 bp upstream of their transcription start sites. Thus, for GAL4 protein, as for λ repressor (Hochschild et al. 1983), the DNA-binding function can be separated from the positive control function.

Repression by a Bacterial Repressor Protein in Yeast
The *E. coli* LexA protein, synthesized in yeast, can bind to its operator and repress *GAL1* transcription if the operator is placed between UAS$_G$ and the *GAL1* TATA homology (Brent and Ptashne 1984).

REFERENCES
Brent, R. and M. Ptashne. 1984. A bacterial repressor protein or a yeast transcriptional terminator can block upstream activation of a yeast gene. *Nature* **312:** 612.
Giniger, E., S. Varnum, and M. Ptashne. 1985. Specific DNA binding of GAL4, a positive regulatory protein of yeast. *Cell* **40:** 767.
Hochschild, A., N. Irwin, and M. Ptashne. 1983. Repressor structure and the mechanism of positive control. *Cell* **32:** 319.

cis and trans-acting Regulatory Elements at the Yeast *HO* Promoter

K. Nasmyth, L. Breeden, and A. Miller

MRC Laboratory of Molecular Biology
Cambridge, CB2 2QH England

All characterized yeast promoters are under positive control in so much as they require not only prokaryotic promoter-like sequences called TATA boxes near the start of transcription but also specific DNA sequences 50–300 bp further upstream. The latter are called upstream activation sequences (UASs) and have many similarities to enhancers in mammalian cells. Most regulation is thought to be positive since in the three best-characterized systems, *GAL* (West et al. 1984), *HIS4* (Hinnebusch et al. 1985), and *CYC1* (Guarente et al. 1984), it is exerted by gene-specific activators acting in *trans* at the UAS. This paper describes a study of the yeast *HO* promoter that suggests that three separate physiological controls are exerted, not via positive control at the UAS, but more probably via negative control sequences situated between the UAS and the TATA box.

The *HO* Promoter

The *HO* gene encodes an endonuclease that initiates mating-type switching in yeast (Kostricken et al. 1983). As part of the complex regulation surrounding mating-type switching, *HO* transcription is subject to at least three forms of regulation. First, it is repressed in **a**/α diploids by the combined action of the *MAT***a**1 and *MAT*α2 gene products (Jensen et al. 1983). Second, in haploid **a** or α cells, it is transcribed in mother cells but not in daughters. Third, it is transcribed only during a brief period in the G_1 phase of the mother cell cycle, at, or soon after, START (Nasmyth 1983).

The *HO* promoter is unusual in that sequences required for its transcription (and regulation) are distributed over at least 1400 bp (see Fig. 1). It is like other yeast promoters, however, in that two sets of sequences are required in *cis* for transcription: a TATA box at −90 and a UAS between −1250 and −1400 (Nasmyth 1985). The DNA between these two regions can be deleted without significantly reducing the level of transcription. All knowledge concern-

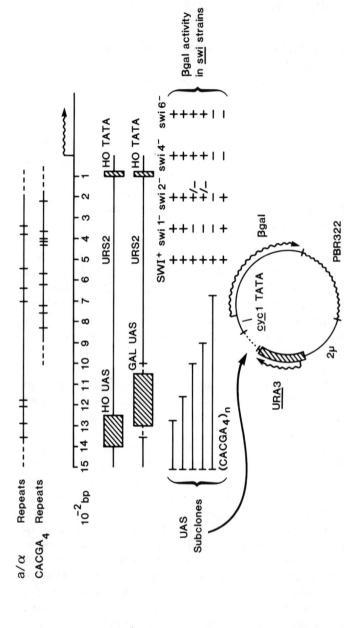

Figure 1 Analysis of the *HO* promoter. The distribution of **a**/α and CACGA₄ control elements within the *HO* promoter are shown at the top of the diagram. The position of the *HO* UAS and its replacement by the *GAL* UAS is shown in the middle of the diagram. Fragments containing different amounts of *HO* promoter DNA were placed next to the *CYC* TATA box in the vector shown at the bottom of the diagram. All fragments had similar UAS activity in the *SWI⁺* strain but variable activities in the various *swi⁻* mutants.

ing *trans*-acting factors has stemmed from genetic studies. Stern et al. (1984) have described five genes called *SWI1,2,3,4*, and *5*, which are required for *HO* transcription. Since then, we have characterized more than 200 new mutants that define four new complementation groups called *SWI6,7,8*, and *9*.

a/α Repression

A deletion analysis of the *HO* promoter has shown that there is no unique region required for **a**/α repression, implying that there may be multiple control sites. We have, therefore, been able to formulate a consensus sequence for the **a**/α control element on the basis of its distribution within the *HO* promoter. The consensus is TC^A_G-TGTNN^A_TNANNTACATCA. There are 9–10 examples scattered throughout the *HO* promoter (Fig. 1) and examples also at other genes known to be repressed by **a***1*/α*2* (*MATα1* and *STE5*). Two 47–50 bp restriction fragments from *HO*, containing different examples of the consensus, can confer **a**/α repression on the *CYC1* promoter when placed between its UAS and TATA box (Miller et al. 1985). Interestingly, the **a**/α control element is less effective at conferring repression when placed upstream of the *CYC UAS* (although there is still a very significant effect) and completely ineffective when placed 30 bp downstream from the start of transcription.

Mother/Daughter Control

The **a**/α control element is a relatively straightforward example of negative control in a eukaryotic cell. The control that ensures that *HO* is expressed in mother cells but not their daughters may be another, but more complicated, example. Our present understanding of this system stems from the analysis of both *cis-* and *trans-*acting mutations. Deletion of all DNA between −150 and −900 leaves an active *HO* promoter that, though altered in its cell-cycle control (see below), is still not transcribed during the G_1 and S phase of the daughter cell cycle. Transcription from this deletion is still dependent upon *SWI1* and possibly also *SWI2* but not upon other *SWI* genes such as *SWI4* and *6*. It seems possible, therefore, that *SWI1* and *2*, rather than *SWI4* or *6*, are involved in the mother/daughter control.

The sequences required for *SWI1* or *2* dependence have been investigated by inserting various fragments containing the *HO* UAS into a version of the *CYC1* promoter that has had its UAS deleted but TATA box retained (Guarente and Maison 1983; see Fig. 1). The fragments all contain the *HO* UAS at their common left-hand endpoint but include varying amounts of the DNA on its 3′ side. All fragments confer a similar upstream activation in SWI^+ strains but only those that include DNA proximal to -1150 show $SWI1$ or $SWI2$ dependence. The observation that a small DNA fragment containing the HO UAS is not $SWI1$ or $SWI2$ dependent and only acquires such dependence upon the addition of further downstream DNA sequences suggests that $SWI1$ and 2 are not positive regulators (in the sense of crp in $E.$ $coli$ or $GAL4$ in yeast) but more likely act to relieve some form of negative control. This hypothesis is consistent also with the observation that transcription from a promoter fusion in which the HO UAS is replaced by the GAL UAS (Fig. 1) is not only galactose dependent but also, partly, $SWI1$ and 2 dependent.

If $SWI1$ and $SWI2$ act by relieving a negative control, then the defect in HO transcription due to $swi1$ or 2 mutations should be suppressed by a mutation of the repressor that $SWI1$ and 2 are proposed to counteract. Thirty-two suppressors of the $swi1-314$ mutation have, therefore, been isolated. All are recessive and fall into a single complementation group that is unlinked to $SWI2$ and that we will for the moment call $SSW21$. (M. Stern et al. [pers. comm.] have previously reported unlinked suppressors of $swi1$ and $swi5$ with which $ssw21$ mutants may be allelic.) $ssw21$ mutants also suppress the $swi1-1$ mutation but not a $swi2-314$, $swi4-100$ double mutant. It is tempting to speculate that $SSW21$ encodes a repressor whose action is overcome only in mother cells through the action of $SWI1$ and $SWI2$.

Cell-cycle Control
The deletion that removes the DNA between -150 and -900 (a region called $URS2$) disrupts the cell-cycle control of HO. Transcription from this deletion is independent of the START in mother cells and no longer requires $SWI4$ or $SWI6$. The simplest interpretation of this result is that $URS2$ confers cell-cycle regulation via a negative control whose repression is normally relieved transiently at START by $SWI4$ and $SWI6$. This hypothesis is consistent with the behavior of a GAL/HO promoter fusion in which the HO UAS is replaced by a 330-bp fragment containing the GAL UAS (Fig. 1). Transcription from this hybrid promoter is not only galactose dependent but also cell-cycle controlled and completely $SWI4$ dependent.

A cis-acting cell-cycle controlling element within $URS2$ that is $SWI4$ and $SWI6$ dependent has also been identified by virtue of its redundancy. There are seven instances of exact matches to the consensus sequence $PyNNPuCACGA_4$, all within the limits of the -150 to -900 interval (Fig. 1). Moreover, a synthetic oligonucleotide with this sequence restores START-dependent transcription when tandem copies are inserted at the breakpoints of the −150 to −900 deletion. Multiple copies of the synthetic CACGA$_4$ repeat activate transcription when inserted next to the *CYC1* TATA box (Fig. 1). This transcription is START dependent and requires the *SWI4* and *SWI6* gene products.

The exact role of the *SWI4,6*-dependent CACGA$_4$ repeat is presently unclear. A simple model would have *SWI4* and *SWI6* transiently inactivating (at or soon after START) a repressor that recognizes the CACGA$_4$ repeat. This model does not explain, however, why the repeat (at least when present in multiple tandem copies) can promote transcription. An alternative model is that two or more separate functions are required for the cell-cycle control exerted by *URS2*. One could be negative, causing repression, and another could be positive, exerting transient START-dependent activation. The CACGA$_4$ repeat may be the latter element, which, at the wild-type locus, serves merely to counteract the repression (i.e., it is insufficient for upstream activation) but when isolated on its own in multiple tandem copies is sufficient for autonomous upstream activation.

CONCLUSIONS

All three forms of *HO* regulation that we have discussed – **a**/α repression, *SWI1,2* dependence, and cell-cycle control – seem separable from upstream activation itself. Negative control, therefore, is implicated but not yet proved in all three cases. It is tempting to speculate that multiple negative controls may be prevalent in the regulation of other eukaryotic genes that are subject to multiple controls on the DNA since such systems can clearly evolve without involving the evolution of complicated interactions between different elements and their different factors, as may be required for interacting positive controls. The requirement that a combination of events be necessary for an active promoter can be easily satisfied if each pathway of negative control, though subject to different physiological parameters (e.g., in the case of *HO*, absence of **a**1 or α2, *SWI1,2* in a mother cell, *SWI4,6* at START), is sufficient to repress transcription (in the absence of derepressing signals).

REFERENCES

Guarente, L., and T. Maison. 1983. Heme regulates transcription of the *CYC1* gene of *S. cerevisiae* via an upstream activation site. *Cell* **32:** 1279.

Guarente, L., B. Lalonde, P. Gifford, and E. Alani. 1984. Distinctly regulated tandem upstream activation sites mediate catabolite repression of the *CYC1* gene of *S. cerevisiae*. *Cell* **36:** 503.

Hinnebusch, A.G., G. Lucchini, and G.R. Fink. 1985. A synthetic *HIS4* regulatory element confers general amino acid control of the cytochrome *C* gene of yeast. *Proc. Natl. Acad. Sci.* **82:** 498.

Jensen, R., G.F. Sprague, and I. Herskowitz. 1983. Regulation of yeast mating-type inter-conversion: Feedback control of *HO* gene expression by the mating type locus. *Proc. Natl. Acad. Sci.* **80:** 3035.

Kostricken, R., J.N. Strathern, A.J.S. Klar, J.B. Hicks, and F. Heffron. 1983. A site-specific endonuclease essential for mating-type switching in *Saccharomyces cerevisiae. Cell* **35:** 167.

Miller, A.M., V.L. MacKay, and K.A. Nasmyth. 1985. Identification and comparison of two sequence elements that confer cell-type-specific transcription in yeast. *Nature* **314:** 598.

Nasmyth, K.A. 1983. Molecular analysis of a cell lineage. *Nature* **302:** 670.

———. 1985. At least 1400 bp of 5′ flanking DNA is required for the correct expression of the *HO* gene in yeast. *Cell* (in press).

Stern, M., R. Jensen, and I. Herskowitz. 1984. Five *SWI* genes are required for expression of the *HO* gene in yeast. *J. Mol. Biol.* **178:** 853.

West, R.W., R.R. Yocum, and M. Ptashne. 1984. *Saccharomyces cerevisiae GAL1-GAL10* divergent promoter region: Location and function of the upstream activation sequence UAS. *Mol. Cell. Biol.* **4:** 2467.

DNA Elements Determining the Differentiated Expression Pattern of *Drosophila* Yolk Protein Genes

P.C. Wensink, M.J. Garabedian, B.M. Shepherd, and M.-C. Hung

Biochemistry Department and Rosenstiel Center, Brandeis University
Waltham, Massachusetts 02254

In this communication we summarize the enhancer-like properties of DNA elements that determine the differentiated expression pattern of *Drosophila* yolk protein (yp) genes. In *D. melanogaster*, transcripts from the three single-copy yp genes (yp1, yp2, and yp3) occur only in females and have been detected only in two tissues: adult fat bodies and ovarian follicle cells (Barnett et al. 1980; Barnett and Wensink 1981; Brennan et al. 1982). This sex-, time-, and tissue-specific control of mRNA concentration is tight. For example, the difference between the steady-state levels of yp mRNA in females and males is at least 5000-fold. The subjects of this communication are the yp1 and yp2 genes. They are divergently transcribed and their capping sites are 1225 bp apart (Hovemann et al. 1981; Hung and Wensink 1981, 1983; see Fig. 1a).

Identification of *cis*-acting DNA elements that determine the in vivo expression pattern of *Drosophila* genes has been made possible by a germ-line transformation method that uses P-element transposons as transformation vectors (Rubin and Spradling 1982; Spradling and Rubin 1982). Our approach to identifying DNA elements necessary for the yp transcription pattern has been to alter a cloned yp1–2 gene region, introduce the altered region into the germ line, and then assay its pattern of transcription. In all cases we examine transcripts from flies that have a stable, single copy of the altered gene region.

SUMMARY OF RESULTS

1. Each gene is controlled by at least two tissue-specific DNA elements. When the yp1 and yp2 genes were introduced into the germ line in the normal genomic configuration (Fig. 1a), their

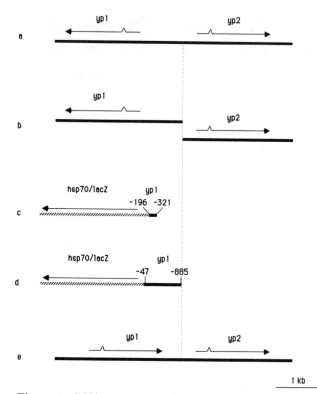

Figure 1 DNA structures introduced into *Drosophila*. The *Drosophila* DNA from the yp gene region is shown by thick, solid lines. The hsp70/*lacZ* gene fusion is shown by thick, hatched lines. Transcribed regions are indicated by arrows that have inverted V symbols showing intron positions. The numbers indicate the positions of breakpoints relative to the capping site of yp1.

transcripts were detected only in the ovaries and fat bodies of the adult female. This is the same pattern of expression as found for the endogenous yp genes. This result indicates that all of the *cis*-acting DNA necessary for the developmental specificities of yp gene expression are contained within this introduced DNA fragment. However, different expression patterns were found when the two genes in this fragment were separated (Fig. 1b) and independently introduced into the germ line (Garabedian et al. 1985). Transcripts from the introduced yp1 gene were detected only in adult female fat bodies, whereas those of the introduced yp2 gene were detected only in adult female ovaries.

Thus, for each gene at least one element permits expression in fat bodies and at least one permits expression in ovaries.

2. A 125-bp fragment located between nucleotides −196 and −321 (relative to the yp1 capping site) controls expression of a heterologous promoter, causing transcripts from the promoter to occur with the same sex, time, and fat body specificities as the yp1 gene in Figure 1b. This occurs when the 125-bp fragment is placed upstream of a previously characterized fusion of the *Drosophila* hsp70 promoter to the *E. coli lacZ* structural gene (Lis et al. 1983). This gene fusion facilitates the assay of gene activity because tissues that have the *lacZ* gene product, β-galactosidase, can be detected by X-gal staining. In our construction (Fig. 1c) the 125-bp fragment is in the same position and orientation relative to the hsp70 promoter as it would normally be relative to the yp1 gene promoter. Since *lacZ* expression was assayed under conditions that do not allow normal transcription from the hsp70 promoter, the expression observed is caused by the addition of the 125-bp fragment and is presumably due to its *cis*-acting influence. To demonstrate that the hsp70 transcriptional control region is not required for this pattern of expression, a second construction was made that replaced the hsp70 upstream region containing the Pelham box (Pelham 1982) and the hstf-binding sites (Parker and Topol 1984; Wu 1984) with yp1 upstream DNA (Fig. 1d). This construction expressed β-galactosidase with the same adult female fat body specificity.

3. The ovarian-specific element near the yp2 gene acts bidirectionally. When the yp1 gene is inverted relative to the yp2 gene and the two genes are introduced into the germ line on the same piece of DNA (Fig. 1e), transcripts from the resulting yp1 gene occur in both fat bodies and ovaries, whereas those of the resulting yp2 gene are found only in the ovaries.

DISCUSSION

These germ-line transformation studies have identified two enhancer-like DNA regions that determine tissue-, time-, and sex-specific gene expression patterns. A 125-bp region can control a heterologous promoter so that it acquires the expression pattern of an adult female fat body. Another region acts in both directions to determine an adult female ovarian expression pattern. These results indicate that both elements can act on both genes, thus supporting the hypothesis that multiple enhancer-like elements with different tissue specificities can influence a single promoter.

REFERENCES

Barnett, T. and P.C. Wensink. 1981. Transcription and translation of yolk protein mRNA in the fat bodies of *Drosophila*. In *Developmental biology using purified genes* (ed. D.D. Brown and C.F. Fox), p. 97. Academic Press, New York.

Barnett, T., C. Pachl, J.P. Gergen, and P.C. Wensink. 1980. The isolation and characterization of *Drosophila* yolk protein genes. *Cell* 21: 729.

Brennan, M.D., A.J. Weiner, T.J. Goralski, and A.P. Mahowald. 1982. The follicle cells are a major site of vitellogenin synthesis in *Drosophila melanogaster*. *Dev. Biol.* 89: 225.

Garabedian, M.J., M.-C. Hung, and P.C Wensink. 1985. Independent control elements that determine yolk protein gene expression in alternative *Drosophila* tissues. *Proc. Natl. Acad. Sci.* 82: 1396.

Hovemann, B., G. Galler, U. Walldorf, H. Kupper, and E.K.F. Bautz. 1981. Vitellogenin in *Drosophila melanogaster*: Sequence of the yolk protein 1 gene and its flanking regions. *Nucleic Acids Res.* 9: 4721.

Hung, M.-C. and P.C. Wensink. 1981. The sequence of the *Drosophila melanogaster* gene for yolk protein 1. *Nucleic Acids Res.* 9: 6407.

———. 1983. Sequence and structure conservation in yolk proteins and their genes. *J. Mol. Biol.* 164: 481.

Lis, J.T., J.A. Simon, and C.A. Sutton. 1983. New heat shock puffs and β-galactosidase activity resulting from transformation of *Drosophila* with an hsp70-*lacZ* hybrid gene. *Cell* 35: 403.

Parker, C.S. and J. Topol. 1984. A *Drosophila* RNA polymerase II transcription factor binds to the regulatory site of an hsp70 gene. *Cell* 37: 273.

Pelham, H.R.B. 1982. A regulatory upstream promoter element in the *Drosophila* hsp70 heat-shock gene. *Cell* 30: 517.

Rubin, G.M. and A.C. Spradling. 1982. Genetic transformation of *Drosophila* using transposable element vectors. *Science* 218: 348.

Spradling, A.C. and G.M. Rubin. 1982. Transposition of cloned P elements into *Drosophila* germline chromosomes. *Science* 218: 341.

Wu, C. 1984. Activating protein factor binds in vitro to upstream control sequences in heat shock gene chromatin. *Nature* 311: 81.

The 68C Glue Gene Cluster
of *Drosophila*

E.M. Meyerowitz, M.A. Crosby, M.D. Garfinkel, C.H. Martin, P.H. Mathers, and K.V. Raghavan

Division of Biology, California Institute of Technology
Pasadena, California 91125

The major synthetic function of the *Drosophila* salivary glands in the third larval instar stage of development is the synthesis of a protein glue. This glue is made in the cytoplasm of the giant salivary gland cells, then secreted to the lumen of the gland. It is expelled from the lumen through the salivary duct at the end of the third larval instar, is deposited on the surface under pupariating larvae, and serves to stick the puparial case to the surface for the duration of the prepupal and pupal periods. The glue consists of seven or eight different polypeptides. Three of these polypeptides, sgs-3, sgs-7, and sgs-8, are translated from messenger RNAs coded in a three-gene cluster located at cytological position 68C4-6 on the left arm of the third chromosome (Meyerowitz and Hogness 1982; Crowley et al. 1983; Garfinkel et al. 1983). This gene cluster is depicted and described in Figure 1. The RNAs coded by the three genes accumulate from early in the third larval instar until a few hours before puparium formation and disappear at the time of pupariation. They are only found in salivary glands and in no other tissue, and they are only present in third instar larvae and in no other developmental stage. At the time when the 68C glue genes are being transcribed, the 68C region of the polytene salivary gland chromosome is prominently puffed. The goal of our work with these genes is to determine the *cis*- and *trans*-acting elements that interact to produce the tissue- and time-specific expression of the 68C glue genes and to find the relation between the 68C chromosome puff and the genes that reside within it.

trans-acting Factors

We know of two factors that act in *trans* in the regulation of the 68C gene cluster. The first is the steroid hormone ecdysterone. This hormone acts both in the initiation of expression of the glue RNAs and in the cessation of this expression. The evidence that ecdysterone

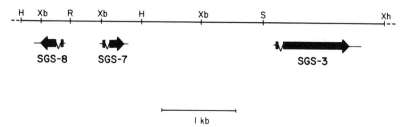

| kb

Figure 1   The 68C glue gene cluster. The three glue polypeptide genes *Sgs-8*, *Sgs-7*, and *Sgs-3* are contained in less than 5 kb of DNA. A restriction map of this DNA is shown, with the following restriction endonuclease cleavage sites; *Hin*dIII (H), *Xba*I (X), *Eco*RI (R), and *Sal*I (S). The directions of transcription of each of the RNAs derived from these genes are indicated by arrows, and the single intervening sequence present near the 5' end of each of the RNAs is indicated by an inverted caret. The thick region of the arrows is the protein-coding sequence in the RNAs; the thin parts are un-translated. The *Sgs-8* RNA is 360 nucleotides in length, the *Sgs-7* RNA is 320 nucleotides, and the *Sgs-3* RNA is approximately 1120 nucleotides; all lengths were calculated without including the poly(A) tail found on each of the transcripts.

is required for the start of 68C glue RNA accumulation comes from experiments with larvae carrying the *l(1)su(f)*^ts67g mutation. This mutation is thought to prevent synthesis of ecdysterone at 30°C but to allow normal levels of the hormone to accumulate at lower temperatures. When larvae are shifted from permissive to restrictive temperatures before the time when glue mRNA synthesis would normally begin, the RNAs do not accumulate even after many hours. If such larvae are fed ecdysterone, 68C RNA accumulation does begin (Hansson and Lambertsson 1983). Little is known about the processes affected by ecdysterone in initiation of 68C RNA accumulation: transcription initiation, elongation, and RNA stability are all possible levels of the effect. The effect of ecdysterone on termination of 68C RNA synthesis when it is already in progress is better understood. It is to cause cessation of new synthesis of all three of the RNAs; the effect is nearly complete in less than 15 min (Crowley and Meyerowitz 1984). At the same time ecdysterone stops new synthesis of the 68C RNA, it causes regression of the 68C puff. This regression occurs in the presence of cycloheximide, indicating a direct action of the hormone on the puff (Ashburner 1974). The puff regression takes approximately 2 hr to be complete, so it is slower than the effect on RNA synthesis (Ashburner et al. 1974; Crowley and Meyerowitz 1984).

The question of how the same hormonal stimulus both starts and stops RNA accumulation from the same genes is open; two possibilites are that different amounts of the hormone are responsible for the different effects, or that the initial hormonal stimulus that starts 68C RNA accumulation also induces other genes whose products cause reversal of the hormonal effect when they have accumulated sufficiently.

One additional *trans*-acting factor required for expression of the 68C RNAs is the product of the X-chromosome gene *l(1)npr-1*. Larvae with mutations at this locus, or with deficiencies that remove the locus, fail to accumulate any of the 68C mRNAs as assayed in RNA blot experiments and fail to synthesize any of them as measured in 15-min pulse-labeling experiments (Crowley et al. 1984). This mutation apparently does not affect ecdysterone concentration (Fristrom et al. 1981).

One curious feature of the phenotype of both *l(1)su(f)*[ts67g] at the restrictive temperature, and *l(1)npr-1* is that the 68C puff is present in larval salivary glands despite the absence of accumulation of transcripts from the genes in the puff (Hansson et al. 1981; Crowley et al. 1984).

cis-acting Factors

Experiments with chromosomal rearrangements show that at most 20 kb of DNA, including the three 68C glue genes, is required in vivo for normal expression of the genes and for puffing of the 68C region when they are expressing (Meyerowitz and Crosby 1982; M.A. Crosby and E.M. Meyerowitz, in prep.). No *cis*-acting sequences farther than 15 kb from any of the genes are required for their control. To identify which sequences are required, pieces of the 68C gene cluster have been inserted into P-element transformation vectors and reintroduced to the *Drosophila* genome. Various constructs (M.A. Crosby and E.M. Meyerowitz, in press.) using pieces of DNA including the *Sgs-3* gene show that no more than 2270 bp of 5′ upstream material are required for normal expression. Removal of sequences from −2270 to −130 leaves an *Sgs-3* gene that expresses at a level approximately 10-fold lower than normal but that expresses in the correct tissue at the correct time. Thus, far-upstream sequences are required for determining the level of *Sgs-3* expression, but only a small amount of 5′ sequence, the gene, and a small extent of 3′ sequence suffice for normal developmental regulation. Similar experiments by Richards and his collaborators (Bourouis and Richards 1985) lead to similar conclusions, but in their hands 130 bp of upstream sequence are insufficient for detectable *Sgs-3* expression. These experiments use an *Sgs-3* gene derived form a different source, different transformation vectors, and different recipient strains than used in our studies; which if any of these elements accounts for the different results is not yet known.

One question raised by the proximity of three coordinately regulated genes is, Do the genes share a single, common set of control sequences, or is each of the genes able to function when separated from the others? *Sgs-3* certainly functions alone, but the *Sgs-3* experiments leave open the possibility that the sequences upstream of that gene are also required for the other two genes in the puff. To find out if this is the case, we have made transformed fly lines containing *Sgs-8* 5′ and 3′ sequences surrounding various gene constructs, with none of the sequences necessary for *Sgs-3* expression included. In these lines the tissue and time of expression of the introduced genes is the same as that of the resident *Sgs-8* genes. Thus, there appear to be at least two independent sets of control sequences in the 68C gene cluster and no single sequence that confers tissue or time specificity in the expression of all of the genes.

ACKNOWLEDGMENTS
The work described here was supported by grant GM-28075 to E.M.M. from the National Institutes of Health. M.D.G. and P.H.M. are supported by National Research Service Award 5 T32 GM-07616; C.H.M. is supported by a National Science Foundation predoctoral fellowship; and K.V.R. is supported by a Procter and Gamble postdoctoral fellowship.

REFERENCES
Ashburner, M. 1974. Sequential gene activation by ecdysone in polytene chromosomes of *Drosophila melangaster*. II. The effects of inhibitors of protein synthesis. *Dev. Biol.* **39:** 141.
Ashburner, M., C. Chihara, P. Meltzer, and G. Richards. 1974. Temporal control of puffing activity in polytene chromosomes. *Cold Spring Harbor Symp. Quant. Biol.* **38:** 655.
Bourouis, M. and G. Richards. 1985. Remote regulatory sequences of the *Drosophila* glue gene *Sgs-3* as revealed by P-element transformation. *Cell* **40:** 349.
Crowley, T.E. and E.M. Meyerowitz. 1984. Steroid regulation of RNAs transcribed from the *Drosophila* 68C polytene chromosome puff. *Dev. Biol.* **102:** 110.

Crowley, T.E., M.W. Bond, and E.M. Meyerowitz. 1983. The structural genes for three *Drosophila* glue proteins reside at a single polytene chromosome puff locus. *Mol. Cell. Biol.* **3:** 623.

Crowley, T.E., P.H. Mathers, and E.M. Meyerowitz. 1984. A *trans*-acting regulatory product necessary for expression of the *Drosophila melanogaster* 68C glue gene cluster. *Cell* **39:** 149.

Fristrom, D.K., E. Fekete, and J.W. Fristrom. 1981. Imaginal disc development in a non-pupariating lethal mutant in *Drosophila melanogaster. Wilhelm Roux's Arch. Dev. Biol.* **190:** 11.

Garfinkel, M.D., R.E. Pruitt, and E.M. Meyerowitz. 1983. DNA sequences, gene regulation and modular protein evolution in the *Drosophila* 68C glue gene cluster. *J. Mol. Biol.* **168:** 765.

Hansson, L. and A. Lambertsson. 1983. The role of su(f) gene function and ecdysterone in transcription of glue polypeptide mRNAs in *Drosophila melanogaster. Mol. Gen. Genet.* **192:** 395.

Hansson, L., K. Lineruth, and A. Lambertsson. 1981. Effects of the *l(1)su(f)*[ts67g] mutation of *Drosophila melanogaster* on glue protein synthesis. *Wilhelm Roux's Arch. Dev. Biol.* **190:** 308.

Meyerowitz, E.M. and M.A. Crosby. 1982. Molecular limits of the 68C glue puff. *Caltech Biology Annual Report*, p. 60.

Meyerowitz, E.M. and D.S. Hogness. 1982. Molecular organization of a *Drosophila* puff site that responds to ecdysone. *Cell* **28:** 165.

Identification of DNA Sequences Required for the Regulation of *Drosophila* Alcohol Dehydrogenase Gene Expression

J.W. Posakony, J.A. Fischer, V. Corbin, and T. Maniatis

Department of Biochemistry and Molecular Biology
Harvard University, Cambridge, Massachusetts 02138

The expression of *Drosophila* alcohol dehydrogenase (*Adh*) genes is regulated at several different levels. The amount of ADH activity varies significantly during *Drosophila* development, and *Adh* expression is limited to specific tissues (Ursprung et al. 1970). In *D. melanogaster* there is one copy of the *Adh* gene per haploid genome (Goldberg 1980), and this gene has two different promoters that are utilized at different times during *Drosophila* development (Benyajati et al. 1983). One promoter ("adult" or distal) is expressed at a low level in embryos and larvae, and at a high level in adults, whereas the other promoter ("larval" or proximal) is expressed primarily in late embryos and larvae (M. Ashburner, pers. comm.).

A variety of methods are available to investigate *Adh* gene expression experimentally, including enzymatic activity measurements, histochemical staining, and RNA analysis. In previous work in this laboratory, we have utilized these techniques to study the expression of an 11.8-kb fragment of the *D. melanogaster Adh* gene (Goldberg 1980), which was introduced into the germ line of ADH null flies by P-element-mediated transformation (Goldberg et al. 1983). The expression of the transduced *Adh* gene was found to be quantitatively and qualitatively indistinguishable from that of the endogenous *Adh* gene.

We have since taken two approaches to the identification of *cis*-acting DNA sequences required for the normal pattern of *Adh* gene expression. In the first approach we have introduced deletions into the cloned *D. melanogaster Adh* gene and then determined the effect of these mutations on *Adh* expression in larvae and adults using P-element transformation. These studies have identified three distinct regulatory elements. In the second approach we have analyzed the expression of *Adh* genes from *D. mulleri*, a member of a species complex in which the *Adh* gene has been duplicated (Barker and Mulley 1976; Batterham et al. 1983; D. Sullivan, pers. comm.). In contrast to the *D. melanogaster Adh* system, where two different promoters are utilized at different stages of development, *D. mulleri* has two different *Adh* genes, one of which is expressed only in larvae, and the other of which is expressed in both larvae and adults (a pattern similar to the differential utilization of the two promoters in *D. melanogaster*). A comparison of the larval to adult switch in *Adh* gene expression in these two species should identify common regulatory components and possibly reveal changes in gene regulation that have occurred during evolution.

Analysis of *D. melanogaster Adh* Gene Expression

As described above, all of the DNA sequences required for normal *Adh* gene expression in P-element transformation experiments are contained within an 11.8-kb DNA fragment that includes 5.5 kb and 4.5 kb of 5'- and 3'-flanking sequences, respectively (Goldberg et al. 1983). To localize the *Adh* regulatory sequences within this fragment, we have analyzed the effect of a series of 5' and 3' deletions on *Adh* gene expression (Fig. 1). When the 4.8-kb *Eco*RI or 3.2-kb *Xba*I fragments of the gene are inserted into a P-element vector and introduced into ADH null flies, approximately normal levels of ADH activity and mRNA are observed in adults. Therefore, all of the sequences necessary for ADH gene expression in adults are contained within the *Xba*I DNA fragment, which includes 0.66 kb and 0.64 kb of 5'- and 3'-flanking sequences, respectively. In contrast, the *Eco*RI and *Xba*I fragments exhibit significantly reduced *Adh* expression in larvae, though the appropriate larval tissue specificity is retained. Thus, deletion of DNA sequences 5' to the *Eco*RI site reduces larval *Adh* gene expression but has little or no effect on adult expression. We conclude that a regulatory element necessary for high levels of larval *Adh* gene expression is located at least 2 kb upstream from the larval promoter. Although we have not yet localized this sequence, we have shown that its activity is independent of its orientation with respect to the *Adh* gene.

When the DNA sequences between the *Xba*I (−660) and *Sal*I (−69) sites are deleted (Fig. 1), detectable activity of the adult promoter is lost, but the larval promoter continues to express (in larvae) at the low level observed with the *Xba*I fragments. We conclude that sequences between −660 and −69 upstream of the distal promoter are required for normal *Adh* expression in adults. This

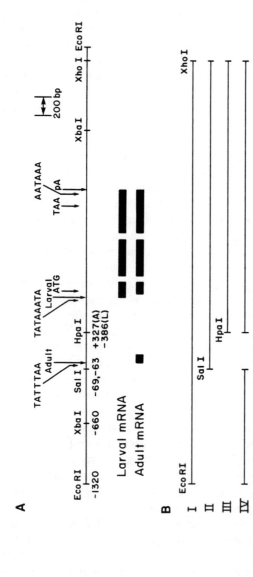

Figure 1 A restriction map of the *D. melanogaster Adh* gene, showing the location of deletions used to identify regulatory sequences. (*A*) A map of restriction sites adjacent to and within the *Adh* gene. Numbers below the line indicate the number of base pairs upstream (−) or downstream (+) from the adult and larval transcription start sites, which are indicated by the arrows labeled Adult or Larval. The initiation and termination sites of translation are indicted by ATG and TAA, respectively. The adult and larval TATA box sequences are indicted by the arrows labeled TATTTAA and TATAAATA, respectively. The arrows labeled AATAAA and pA indicate the polyadenylation signal and polyadenylation site, respectively. The exons present in larval and adult mRNA are indicated by the black boxes below the line, while the spaces between these boxes represent introns. (*B*) A map of deletion mutants discussed in the text.

region must contain an adult regulatory and promoter sequence. Truncation of the gene to the *Hpa*I site (Fig. 1), which lies 387 bp 5' to the larval promoter transcription start, preserves a similar low level of larval promoter activity. Moreover, this expression exhibits appropriate tissue specificity in larvae, indicating that the regulatory elements responsible are located 3' to the *Hpa*I site.

Evidence for an adult regulatory sequence that is separable from the adult promoter is provided by an analysis of the *Sal*I/*Hpa*I internal deletion, which removes a 390-bp fragment that includes the adult TATA box and transcription startpoint. With this deletion, a high level of *Adh* gene expression is observed in adult flies, but the transcripts are accurately initiated at the larval promoter. Thus, there appears to be an adult upstream regulatory element that can act on the larval promoter in adults when the adult promoter is deleted. Remarkably, high levels of transcripts initiated at the larval promoter are also observed in larvae with the *Sal*I/*Hpa*I deletion. This observation suggests that the temporal and/or quantitative activity of the adult upstream regulatory element can be altered when it is placed directly adjacent to the larval promoter and its associated regulatory sequences.

Evidence that the adult upstream sequence has the properties of a regulatory enhancer element has been obtained by analyzing P-element-transformed lines carrying a construction in which the region between the *Eco*RI and *Sal*I sites 5' to the distal promoter of *Adh* (– 1320 to – 69) are fused to the dopa decarboxylase (*Ddc*) gene at a position 210 nucleotides upstream of its transcription startpoint. The – 210 truncation of *Ddc* exhibits essentially normal expression in transformed flies (J. Hirsh, pers. comm.). Transformants carrying the *Adh-Ddc* fusion, by contrast, have 5-fold to 10-fold elevated levels of DDC activity in adults. This effect of the *Adh* adult upstream sequences on *Ddc* is very similar to their effect on the larval promoter of *Adh*, described above. Thus, these sequences are capable of activating a high level of expression in adult flies, even from a heterologous promoter. Moreover, the *Adh* sequences appear to contain a *cis*-acting element involved in controlling the tissue specificity of *Adh* expression. In the transformed lines carrying the *Adh-Ddc* fusion, DDC activity is easily detectable in the accessory genital apparatus of adult males, a tissue that normally expresses ADH but not DDC.

On the basis of these observations, we propose that *D. melanogaster Adh* gene expression is regulated by at least three distinct regulatory elements. One element is located more than 2000 bp upstream from the larval promoter and controls the level of transcription in larvae. A second element is located within 400 bp of the larval transcription start site. This sequence is necessary for low levels of tissue-specific expression from the larval promoter. The third element is located between −69 and −660 from the adult transcription startpoint. This element can confer temporal and tissue-specific expression on the *Ddc* gene as well as on the larval promoter.

Analysis of *D. mulleri Adh* Gene Expression

The *D. mulleri Adh* genes were isolated from a bacteriophage λ library of *D. mulleri* genomic DNA, using the cloned *D. melanogaster* gene as a hybridization probe. Restriction endonuclease cleavage mapping and DNA sequence analyses revealed the presence of three *Adh* genes per haploid genome. As shown in Figure 2, these genes are located in a closely linked cluster in the order ψ-adult (an adult pseudogene), adult, and larval. The intron/exon arrangement of these genes is identical with that of the *D. melanogaster* genes, but in contrast to the single *D. melanogaster Adh* gene, the *D. mulleri* genes each have a single TATA box and a single startpoint of tran-

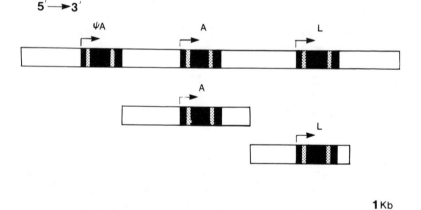

Figure 2 A diagram showing the organization of the *D. mulleri Adh* genes. The three *Adh* genes are indicated by the black boxes labeled ψA (adult pseudogene), A (adult gene), and L (larval gene). The black, stippled, and white boxes represent exons, introns, and flanking DNA sequences, respectively. Arrows indicate the transcription startpoints of the genes, while the arrow labeled 5′→3′ indicates the direction of transcription of all three genes. The adult and larval genes shown below the gene cluster represent the extent of flanking sequences present in the single-gene transformants discussed in the text.

198

scription. Analysis of *Adh* transcription in *D. mulleri*, using adult or larval gene-specific hybridization probes, indicates that the pattern of *D. mulleri* larval and adult *Adh* gene expression parallels the expression from the larval and adult promoters of *D. melanogaster*. In addition the pseudogene, which contains numerous deletions and insertions in the coding region, is transcribed at low levels in adults. Moreover, with the exception of two adult tissues, the pattern of tissue-specific expression of the *Adh* genes is the same in the two species. The primary difference between *Adh* gene expression in the two species is quantitative. The *D. mulleri* adult gene is expressed at high levels in larvae, while the *D. melanogaster* adult promoter is expressed at low levels at the corresponding stage of development.

To determine whether the *D. melanogaster* regulatory factors can recognize the *D. mulleri* control sequences, the entire *Adh* gene cluster from this species was introduced into the germ line of *D. melanogaster* by P-element transformation. Remarkably, the temporal and tissue-specific expression of the *D. mulleri* genes in the transformants is appropriately regulated. However, the quantitative distribution of adult and larval gene transcripts is like *D. melanogaster* rather than *D. mulleri* (i.e., there is relatively little adult transcript in larvae). We conclude that the putative *trans*-acting factors that regulate the larval to adult promoter switch in *D. melanogaster* can act on corresponding sequences in the adult and larval *D. mulleri* genes. Moreover, the quantitative pattern of the adult *D. mulleri* gene expression observed in the transformants appears to be governed by the *trans*-acting factors in *D. melanogaster*.

To determine whether the *D. mulleri* adult and larval genes can be regulated autonomously, DNA fragments containing only one of the two genes (Fig. 2) were separately introduced into *D. melanogaster* by P-element transformation. Transcriptional analysis of the transformants indicated that the larval gene alone is appropriately regulated. The larval gene is not expressed in adults, and its quantitative and tissue-specific expression is normal. Thus, the DNA sequences within and immediately flanking the larval gene are sufficient for normal larval *Adh* gene expression. A different result was obtained when transformants containing only the adult gene were analyzed. In this case the temporal expression of the gene is correct, but the level of transcription is at least 20-fold less at every stage of development. Thus, although the elements necessary for adult gene expression appear to be within or near the gene, the sequences that regulate the level of transcription must be located upstream or downstream from the fragment used in the transformation experiment. Experiments are in progress to localize this sequence and the sequences that determine the stage- and tissue-specific expression of the two *D. mulleri* genes. A comparative study of the *Adh* genes from the species with single and duplicated *Adh* genes should provide insights into the mechanism of the larval to adult switch in *Adh* gene expression.

ACKNOWLEDGMENT

This work was supported by a grant from the National Institutes of Health to T.M.

REFERENCES

Barker, J.S.F. and J.C. Mulley. 1976. Isozyme variability in natural populations of *Drosophila buzzatii*. *Evolution* **30:** 213.

Batterham, P., J.A. Lovette, W.T. Starmer, and D.T. Sullivan. 1983. Differential regulation of duplicate alcohol dehydrogenase genes in *Drosophila mojavensis*. *Dev. Biol.* **96:** 346.

Benyajati, C., N. Spoerel, H. Haymerle, and M. Ashburner. 1983. The messenger RNA for alcohol dehydrogenase in *Drosophila melanogaster* differs in its 5' end in different developmental stages. *Cell* **33:** 125.

Goldberg, D.A. 1980. Isolation and partial characterization of the *Drosophila* alcohol dehydrogenase gene. *Proc. Natl. Acad. Sci.* **77:** 5794.

Goldberg, D.A., J.W. Posakony, and T. Maniatis. 1983. Correct developmental expression of a cloned alcohol dehydrogenase gene transduced into the *Drosophila* germ line. *Cell* **34:** 59.

Ursprung, H., W. Sofer, and N. Burroughs. 1970. Ontogeny and tissue distribution of alcohol dehydrogenase in *Drosophila melanogaster*. *Wilhelm Roux's Arch. Dev. Biol.* **164:** 201.